Testimonials

"Intelligent automation has emerged as a critical discipline for businesses seeking to improve operational efficiency and reduce toil. Co-authored by industry and academic experts, this book promises to offer practical guidance and valuable insights rooted in real-world experience and cutting-edge research. It covers topics such as citizen-led development initiatives, the strategic use of automation platforms, and the role of AI and ML, including the latest Generative AI technologies. Additionally, it provides actionable advice on aligning automation initiatives with organizational culture and business objectives. This book is a valuable resource for anyone aspiring to become a leader in the field of Intelligent Automation."

Rama Akkiraju, *VP, Enterprise Automation and AI, NVIDIA*

"With a profound historical understanding and unparalleled insight into AI and intelligent automation, Shail Khiyara is a true thought leader. The chapters contributed by Shail in this book offer a masterful blend of in-depth knowledge, historical context, and forward-thinking insight. As a speaker, Shail is engaging and enlightening, transforming complex concepts into relatable narratives."

Darren Atkins, *CIO, The Royal Free London NHS (UK National Health Service)*

"With automation being such a critical and transformative component of the future of business, this book could not have come at a better time. There's a fundamental shortage of talent of those who deeply understand automation capabilities and how to appropriately apply them to business solutions. This book is a tremendous resource for anyone who is looking to upskill, reskill, or simply learn more about intelligent automation."

Wendy Batchelder, *SVP & Chief Data Officer, Salesforce*

"I've been in the intelligent automation space for almost 6 years and have had the opportunity to work with many of the Fortune 500. We've partnered with them to operationalize large-scale intelligent automation programs that deliver hard dollars back to their balance sheet and transformed their business processes for the future of work.

"The release of this book is very timely; many organizations struggle with successfully scaling their IA programs. With the acceleration of AI, this will only become more complex, and organizations that have an automation-first mindset will set themselves apart from their competition. This is why bridging the gap between business and academia is crucial in ensuring the future talent of work supports this demand."

Emmanuel R. Galan, *Regional Vice President – MEU West, UiPath*

"This book is a must-read for EDUCATORS AND EXECUTIVES ALIKE who have an interest in learning more about Automation and how it will affect businesses and the global economy. There is no doubt that RPA and intelligent automation have gathered pace and steam over the last few years, and this book captures the opportunity from a practical perspective. The contributors to this book have real firsthand experience at implementation and have shared their candid journeys and experiences. I hope you enjoy this as much as I did."

Ray Grady, *former CEO Conexiom*

"Companies today all know they have to automate. They have to improve cost measurement, speed, productivity, and go-to-market strategies – while freeing employees from mundane, repetitive tasks. It's all about combining the best of human creativity and the power of digital technology. Embrace it… as it's going to happen, and no better book to tell you how!"

Adrian Jones, *CEO, and Co-Founder, personar.ai*

"The subject matter, Robotic Process Automation and Intelligent Automation, of this book is very timely. The authors provide a very readable book for both practitioners and academicians. The chapters are broken down such that one can choose a specific chapter that is more applicable to one's area of interest. The authors present numerous topics that organizations can implement to remain competitive. This book has something to offer for everyone."

Dr. Gerald Kohers, *Department Chair - Management, Marketing, and Information Systems, Sam Houston State University*

"In my Celonis experience working with Bobby Jutley on Process Mining opportunities at HP, I have observed that Bobby is a highly effective change agent in digital transformation and Automation for the Company. He is quick to identify and promote opportunities, and his strong ability to collaborate across IT and the business sets him apart from most people in the business process automation space. His attention to detail and sense of urgency is key to transforming insights to value for HP."

Susan Lucero, *Global Account Manager, Celonis*

"With our IA journey, we learned it is important to integrate process discovery, mining, and automation to solve the right problem. Technology is becoming more simpler (with little or no coding), which is so important to have robust planning: from business drivers to governance strategy, in order to achieve the organization's digital transformation outcomes. This book bridges theory with real examples to help practitioners plan their organization's IA strategy."

Emilie Ly, *Senior Director, BPM & RPA, VMware*

"Intelligent automation, and especially Robotic Process Automation, requires building a bridge between business needs and the application of technology and academic thinking to solve tomorrow's challenges today. This book provides a clear and concise view of how to learn and adopt Intelligent Automation and expands our vision of how this new industrial revolution is already impacting our schools and businesses. The success of any company is always marked by creating competitive advantages. Intelligent automation provides a gateway to create value within our community, customer base, and future workforce."

Antonio Marin - *CIO, USMed-Equip, LLC.*

"Gigaforce clients were skeptical about implementing RPA in their Subrogation and Recovery department, but it has exceeded their expectations. Implementing intelligent automation for business processes has been a game-changer. We have saved time, reduced errors and costs, and improved our client's overall efficiency. Our client team members can now focus on more strategic tasks requiring human expertise in recovery and subrogation. This book presents how to establish, scale, and manage intelligent automation practices."

Rajeev Rawat, *Co-Founder, and CPO, Gigaforce Inc.*

"The authors do an awesome job of describing in this book how Intelligent Automation is the next evolutionary step for achieving sustainable competitive advantage! Intelligent Automation integrates Robotic Process Automation (RPA), Artificial Intelligence (AI), and Machine Learning (ML) by automating business workflows to advance business processes. This book is a fantastic reference for academics, students, and researchers, but it provides an even stronger guide for the practitioner trying to achieve and sustain a competitive advantage in today's global automated business world."

Dr. Gary L Stading, *Professor of Supply Chain Management, Texas A&M University - Texarkana*

"Bobby Jutley brings his expertise and real-world experience to light as he succinctly summarizes the technological and organizational aspects of Process Discovery while highlighting the fact that Process Discovery excellence has rapidly become a strategic weapon and critical success factor in prioritizing digital transformation opportunities and creating competitive advantage by driving quality, speed, and agility into every aspect of the business."

Ramey Stevens, *Global Account Executive, UiPath*

"RPA has been one of the most critical capabilities in his technology toolbox. Paul says that without RPA, "most CIOs and business leaders simply would not have been able to advance digital transformation at the pace necessitated. Digital transformation requires a comprehensive suite of tools and capabilities, and RPA fits in perfectly."

Paul Walsh, *SVP of Digital Technology and Operations at Sony Interactive Entertainment*

Intelligent Automation

Since prehistoric times, humans have invented ways to simplify daily activities to improve productivity. The most recent milestone in this journey is robotic process automation (RPA), helping to build software robots that can be leveraged to automate mundane and repetitive tasks that can be labor-intensive and prone to errors. In recent years, RPA has been integrated with emerging artificial intelligence (AI) and machine learning (ML) technologies to create what is referred to as intelligent automation (IA), emulating human actions and decision-making abilities. This book addresses the critical questions about the rise, usage, and future of IA practices.

This book is structured by general personas considered as its primary target audience, ranging from:

- Early-stage practitioners seeking to learn effective management of IA programs

- Established IA practitioners seeking to drive maturity and scale

- Business leaders seeking to understand how to drive business value using IA

- Practitioners or academicians seeking to collaborate

This book is strongly recommended for practitioners seeking to plan, implement, and scale IA practices in their organization and for researchers and students who intend to study strategy, implementation, and management of IA practice to accelerate the digital transformation agenda.

Intelligent Automation
Bridging the Gap between Business and Academia

Edited by
Marie Myers, Carol Brace, and Lila Carden

CRC Press
Taylor & Francis Group
Boca Raton London New York

CRC Press is an imprint of the
Taylor & Francis Group, an **informa** business
A CHAPMAN & HALL BOOK

Cover design inspired by Chat GPT.

First Edition published 2024
by CRC Press
2385 NW Executive Center Drive, Suite 320, Boca Raton FL 33431

and by CRC Press
4 Park Square, Milton Park, Abingdon, Oxon, OX14 4RN

CRC Press is an imprint of Taylor & Francis Group, LLC

ISBN: 978-1-032-23175-4 (hbk)
ISBN: 978-1-032-21806-9 (pbk)
ISBN: 978-1-003-27612-8 (ebk)

DOI: 10.1201/9781003276128

Typeset in Minion
by codeMantra

Contents

Author Testimonials

"As digital transformation progresses rapidly in every aspect of modern life, this book provides a step-by-step guide to turning academic and theoretical concepts on intelligent automation into implementable practice across organizations. The authors have shared their wide-ranging industry and academic experience in all aspects, from designing programs to upskilling talent to delivering sustainable value, to provide an essential playbook for every real-world practitioner on the intelligent automation journey."

Aftab Ahmed, *Manager, Compliance & IT Audit, ConocoPhillips*

"If you want to exercise your editorial "pen" to shorten and sweeten it, I would be most delighted !!!" This book is one of the rare occasions where so many industry experts from business and academia converge to bring together a very thought-provoking point of view to guide the audience through a journey of the past, present, and future of intelligent automation and future. And more importantly, this book articulates time-tested and practical steps, methods, and considerations to successfully establish and adopt intelligent automation practices to drive meaningful business transformation."

Partha Baral, *Head of Digital Process & Automation at HP*

"As technology continues to evolve at such a fast pace, and we continue to ride the wave of the 4th Industrial Revolution, this book provides an excellent context for how academia can be bridged to industry practice, with the underpinning of insightful and robust commentary from leading practitioners sharing their digital transformation experience. The book talks about practical steps to stand up a CoE and what ingredients are needed to scale. The narrative clearly calls out the importance of implementing digital transformation and the need to weave in a construct that enables the workforce to embrace digital skills and prepare for the future of work. This book provides a comprehensive read as we move towards AI-powered automation and endless possibilities."

Bobby Jutley, *Interim Head of Digital Process and Automation at HP*

"The future of work is rapidly evolving, driven by new digital ways of working, including intelligent automation. It is imperative for students to adapt and equip themselves with the necessary skills to thrive in this dynamic landscape. By embracing emerging technologies and fostering a mindset of continuous learning, students can unlock endless opportunities and shape a successful career in the digital era."

Kriti Kapoor, *Global Director, Digital Practices & Communities, HP Inc.*

"Leaders must be educated for business and society - often combining seemingly incompatible goals. I am thrilled to have contributed to this symbiosis that bridges the changing business landscape of Automation and AI with academia. Digital transformation is also about the transformation of people. This book illuminates a wide spectrum of boundless possibilities shared by renowned experts."

Shail Khiyara, *President & COO, Turbotic - Founder, VOCAL Council*

"This book is a timely reminder for higher education about the growth in AI and the need to partner with companies to ensure that students are well equipped for the future of work."

Dr. Tiffany Maldonado, *Management, Marketing, and Information Systems, Sam Houston State University*

"I'm passionate about contributing to the efforts to increase student's learning opportunities and enhance their academic achievements. Having worked in the intelligent automation space for over 5 years, enabling some of the world's largest organizations to scale through their Intelligent Automation journey, I was thrilled with the opportunity to contribute to this book to connect my two passions together. With the growing pace and complexity of AI, this book is well-timed for practitioners and students alike, presenting established best practices for implementing and scaling the IA practices."

Neeraj Mathur, *Director, Intelligent Automation, VMware.*

Foreword

THROUGH THE LAST THREE centuries, society has increased global domestic productivity at a blindingly fast pace. This rapid evolution has been greater than the world has ever experienced before the industrial age and the currently unfolding digital transformation era. Up until the mid-20th century, automation was dependent on analog mechanics and structures to increase productivity, which effectively minimized the ability to work at a global scale efficiently. In the current era, coined as "Industry 4.0", the world is experiencing the next steps in automation through digitized "smartening" of these devices and creating cyber-physical systems (CPS) that embrace powers such as pattern recognition through machine learning techniques. These techniques, combined with robotic process automation (RPA), harness the powers of intelligent automation (IA) and have led to new abilities of machines to learn, adapt, create, and solve without the interventions of a human.

These digital technologies were never possible in prior industry evolutions that were dependent solely on analog states. The invention of the modern-day transistor and increasing capacities through "Moore's Law" have allowed, about every 18 months, the doubling of capacities of the computer chips empowering digital processing, networking, and storage of data that digitizes the world around us. On the same note, humans have wondered about and even feared the power of automation and its power to disrupt the workforce, which has accelerated into prominence through the current transformation. However, as IA matures, it will be viewed as one of the positive differentiators in determining the tasks handled by humans versus their robotic counterparts. By allowing humans to harness the powers not yet automatable by robotics and allowing robotics to handle the rest, we will see increases in quality and service speed, reduction in error, higher accessibility, lower costs, and value added to the workplace while synergistically creating a happier workforce.

In 2021, I was thrilled to have met Neeraj Mathur at Carnegie Mellon University, who had taken my class on Managing Digital Business while doing his master's studies here. This was also not the easiest time for studies, as we were experiencing the second year and fifth wave of a deadly coronavirus pandemic. When I learned that he was working with several other authors on a book covering IA, I was excited to hear this and was honored to be asked to contribute to the foreword for this book. With his background in IA and RPA, and experiences applying these techniques in the industry, it was an enlightening experience for his peers to have another aspect of this unfolding and emergent field. It is also quite encouraging that he has taken the next step in applying this knowledge to provide you, the reader, with a similar experience through this text. This is a wonderful example of the strength created between the industry and academic partnerships to further the field of IA.

One of the compelling issues moving into the next decade of digital transformation is the talent gaps experienced in IA. Many case studies exist that exemplify the need for internal incubators of IA and related or synergistic technologies. The problems companies experience in moving digital transformations forward are generally characterized by a lack of what Lockheed Martin would refer to as the "skunk works" or "black box" of the organization. These R&D "think tanks," generally kept in a silo away from the core business, either do not exist or experience limited funding due to the lack of manpower, lack of capacity to hire researchers, lack of internal skills, fears of disruption to the core of the business, or a general misunderstanding of the importance of such technologies. As previously mentioned, industry leaders can work to resolve these issues by forging collaborations between their companies and academic institutions. By aligning with the academic institution's research capabilities, both in building on the knowledge and expertise of the practitioners as well as the students who are seeking experiential learning, companies generally see major gains in all of these areas and can achieve far greater results than in just working at it independently. Academic institutions also reap benefits from these partnerships through the real-world exercises that they can apply IA expertise across, while the company gains the benefits of learning about cutting-edge IA research and how to apply it. As these partnerships grow, the benefits usually reach far beyond the initial interests and tend to lead to increased cross-collaborations across other academic departments and disciplines that a company may not have otherwise been able to make initially.

There are many books available on RPA and IA in the market, but it's important to note that this book is designed as a single reference point for two main audiences: those who want to learn more about these practices and those who want to learn how these technologies are developing in the current and future era of Industry 4.0. However, from a third vantage point, this book is also relevant for those already "in the know" of IA and want to experience it through a unique lens which this book will provide. Part of this uniqueness is in its application to multiple industry sectors and how IA fits the mold for everything between aerospace and agriculture. As a result, efficiencies are continually uncovered, and innovations increase through the application of IA. Given the results from the past decade, it's no wonder that this area is one of the hottest trending technologies on the minds of industry leaders, from the tech lean to the tech savvy alike.

To tell this story and to help the reader construct an approach to IA, this book has been organized into ten chapters.

Chapter 1 provides a holistic overview of the power of RPA and its integration into the movement toward "intelligent automation," that is, the practice of increasing units of productivity while reducing units of input through software-based "bots" which can handle tasks.

Chapter 2 explores the roots of RPA and IA success through business process management and process mining techniques. For digital transformations to succeed, they have to start with those directly involved with the day-to-day operations. In many cases, these processes will involve patterns not documented or are not a part of the current digital process. The reader will have the tools necessary to deep dive into a process or function and be able to determine the cost and overall effectiveness if the process were to be automated. This allows for successful implementations, changes to an existing process, or even abandoning a transformation if the knowledge-gathering process proves the transformation is too costly or convoluted to pursue.

Chapter 3 covers how, as digital transformations continue to mold and shape businesses, human capital investments to increase digital aptitude through lifelong learning practices become even more critical to its success. This chapter will explore ways to approach this within the organization as well as through partnerships with academic research institutions for continuing education through the lifecycles that businesses face.

Chapter 4 covers an important and potentially lesser-known feature of IA about its accessibility and applicability. In most cases, getting started in RPA/IA practice does not require extensive programming knowledge, and many IA companies offer "out of the box" solutions that can provide a simple start to the automation process. The reader will be provided with applicable examples to test and showcase IA implementations.

Chapter 5 stresses the importance of understanding cultural and organizational change as it pertains to the automation of tasks. This chapter explores the author's "humble organizational culture" to encourage traits that should be considered when disrupting traditional workflows within an organization.

Chapter 6 will help you understand the IA program's scope, implementation, and best practices. After reading this chapter, you will have a roadmap for implementation and awareness of the application implications through lessons learned from previous implementations.

Chapter 7 covers an all important component of digital technologies, which is the ability to scale up or down depending on the needs of the business. Readers will be given tools to explore different ways of scaling across organizations within the business or across them and some example best practices to consider as you build your roadmap to successful implementation.

Chapter 8 covers the changes in how humans apply effort through the changing workforce and the "future of work." For businesses to stay relevant, it is important that employees selected for the various roles important for future success have the necessary digital skills to apply to the current and the future of the business.

Chapters 9 and 10 cover two key aspects of digital transformation: the types of disruptions that might occur with automation and what types of future disruptions could be waiting for them through developments in RPA and IA technologies. There is no crystal ball or definitive roadmap to the future, but readers will be empowered to understand how to be ready for and implement future IA when disruptive innovations threaten the core of the business and the livelihoods of their employees.

In conclusion, we as a society are facing new and unique challenges in socioeconomic, environmental, and sustainable practices. As we continue to experience and learn from our newly digitized and intelligently automated world, leaders are ushered into the importance of IA's impacts on these areas and must incorporate them into their business. Two high-level phases of IA can provide the tools and techniques. One phase focuses on the economics of the business by giving insight into developing new methods of creating and sustaining competitive advantage against its rivals, while the second phase focuses on empowering new sources and avenues for the creative, brainstorming nature of humans to tackle newly emergent problems, as well as on developing innovations that will solve these problems. Having worked, taught, and experienced the power of these and other digital technologies over the past three decades and across my experiences applying these technologies in numerous industries, including healthcare, fashion, architecture, and education, I can attest that this book will prove to be yet another useful tool that a diverse set of readership that includes students, researchers, practitioners, or those simply inquiring into the interests and power of IA. All will surely benefit from the knowledge gathered from the text within.

I do hope that you will enjoy taking this journey and using this book as a guide to the amazing opportunities that await you in the next generation of digital transformation!

J. David Riel
Distinguished Service Professor of
Digital Business & Digital Transformation
Heinz College of Information Systems & Public Policy
Carnegie Mellon University, Pittsburgh, PA

Acknowledgments

WE WANT TO THANK and acknowledge the following people who supported and provided critical advice and informed perspective on the impacts of Intelligent Automation. Their drive and passion for ensuring we bridge the gap between corporate and academia were insightful and invaluable.

We would also like to thank our employers and families for supporting us while we compiled this work.

This book showcases the personas road map the authors have gone through to support bridging this gap. It is a must-read.

Courtney Banks
Graphic Designer, University of Houston Downtown

Jessica Chen
Chief Administrative Officer, Finance
Hewlett Packard

Emilie Ly
Senior Director, BPM and RPA
VMware Inc.

Antonio Marin
Chief Information Officer
US Med-Equip, Inc.

Julio Viquez
Senior Manager, Intelligent Automation
VMware Inc.

Editors

Marie Myers is the Chief Financial Officer (CFO) for HP Inc. In this role, she is responsible for financial operations including accounting, financial planning and analysis, business decision support, tax, audit, treasury, and investor engagement. Prior to being named CFO in February 2021, Marie served as HP's Chief Transformation Officer (CTO) where she led the company's IT and transformation organizations. Before rejoining HP in 2020, she was the CFO at a robotic process automation company.

With almost 25 years at HP, Marie has held numerous leadership roles in Finance and played a crucial role as Finance Lead for the separation of HP Inc., the world's largest corporate split.

Marie serves on the Board of Directors of F5 Networks, a global company that specializes in cloud and security application services and on the Board of KLA, a semiconductor company. She is also committed to education and to helping ensure STEM for girls and advancing the digital agenda in education, serving on the boards of the University of St. Thomas in Houston and the University of Queensland. Marie is also a member of the National Association of Corporate Directors (NACD) and Chief, a private network built to drive more women into positions of power and keep them there.

Carol Brace is a visionary strategist/analyst with a demonstrated ability to deliver corporate objectives. She has a solid 20-plus-year career that includes developing strong customer relationships, creating market advantages, reducing and controlling expenses, and fostering a culture of teamwork. She is a specialist in planning and implementing processes to improve the effectiveness of business teams and business systems.

She possesses strong analytical and problem-solving skills and the ability to quickly determine how a process or system works, recognize improvements, and envision any new procedures required. She is certified in Lean Six Sigma, Project Management, Supply Chain, and International Logistics. This, coupled with her practicum experience, is a win-win for academia.

In addition, she has a penchant for community service. Touched by the needs of the thousands of displaced workers that were affected by layoffs due to the retirement of the Space Shuttle program (2011), Carol Brace reached out to provide assistance and training for those transitioning into new occupations. She initiated and customized a Lean Six Sigma Green Belt, Lean Six Sigma Black Belt, Project Management Professional (PMP), and Logistics program bringing in leaders from various industries and creating a partnership with Mayor Annise Parker's office. Carol provided the students who participated in the training with the University of Houston valuable opportunities to meet and network with leaders of growing industries as a part of the curriculum. Carol showed her unwavering commitment by commuting hundreds of hours a week and donating her time to help the displaced workers. Her flexibility and wiliness to "make it work" for the students are rarely found. She constantly found ways to provide additional programs and opportunities for the students, never letting an obstacle stand in her way. Carol's diligence and sacrifices have made her an essential part of the community and brought new awareness to the University of Houston Center of Transportation and Logistics training programs. Her passion led to several recognitions: "Top Thirty Influential Woman of Houston," "National Top Ten Business Woman" from the American Business Women Association, "Woman of the Year" from the American Business Women's Association, and "Woman of Excellence" from the Federation of Houston Professional Women.

She is currently sharing her talents with Sam Houston State University, which has provided her with invaluable assistance and support in the writing and editing of this book.

 **Dr. Lila Carden** is an Assistant Professor at University of Houston. She earned her undergraduate degree in Accounting from Texas A&M University, her MBA from the University of Houston, and her PhD from Texas A&M University. She is a Project Management Professional (PMP) certified by the Project Management Institute (PMI).

Prior to becoming a university faculty member, Dr. Carden worked professionally in a number of wide and diverse organizations such as Waste Management; El Paso Energy; Tenneco Gas; Texas General Land Office; and Peat, Marwick, Main & Co. These positions allowed her to gain experiences in the following areas: auditing; oil and gas contracting, scheduling, and nominations; and project and program management. During her tenure in corporate, she developed and coordinated a curriculum for a mentor-mentee program, completed the Project Blueprint Board Leadership Training sponsored by the United Way, and worked as an Adjunct Professor. Carden's passion to advance the preparation of students led her to later teach at four universities in Texas (University of Houston, University of Houston-Downtown, Houston Baptist University, and Texas A&M University) in the following areas: technology project management, risk assessment, quality improvement, human resources, human resource development, and project management. These industry and academic positions have allowed Carden to use theoretical and practical applications of project management and technology techniques to promote collaborations that support outcomes that contribute to the success of academia and industry.

Dr. Carden's research philosophy is to produce a body of knowledge that advances theory and industry practices to prepare students to work in future STEM (science, technology, engineering, and math) professions. She continues to advance her knowledge internationally and regionally by publishing manuscripts in journals such as *Journal of Management & Organization, International Journal of Project Management, Engineering Management Journal, Human Resource Development International, Project Management Journal, International Journal of Six Sigma and Competitive Advantage,* and *Business and Professional Communication Quarterly.*

In closing, Carden's motto is "to live each day uplifting others by advancing educational opportunities and engaging in service for all."

Contributors

 **Aftab Ahmed** is a finance and IT professional with more than 25 years of business experience. He is currently Manager, Compliance and IT/OT Audit at ConocoPhillips, one of the world's leading oil and gas companies, where he has held roles in accounting, financial reporting, strategic planning, risk management, and investor relations and communications. Aftab has worked in business unit and corporate roles, working with a wide range of vendors, consulting organizations, professional and industry groups, regulators, and other external stakeholders.

Aftab was previously Program Manager, Emerging Digital Technologies for Functions, where he led the introduction of intelligent automation into ConocoPhillips through digital transformation efforts across finance and other functional groups, delivering a program from exploration to pilot to scaled adoption across a range of technologies, including robotic process automation, data analytics, and natural language processing. Aftab also worked with industry partners to launch Blockchain for Energy, a collaboration across the oil and gas value chain to identify, test, and deploy use cases that combine blockchain with innovative technologies.

Prior to joining ConocoPhillips, Aftab was Senior Manager at Deloitte in the United Kingdom, with roles in assurance, corporate finance, internal audit, and technology risk consulting. Aftab led audit and advisory engagements with global clients across multiple industries, including agricultural, automotive, information services, manufacturing, and retail.

Aftab is a Chartered Accountant and Certified Internal Auditor. He earned a Bachelor's degree in Money, Banking, and Finance from the University of Birmingham and a Master of Business Administration from Warwick Business School. Aftab currently resides in Houston, Texas.

 **Partha Baral** is a versatile and forward-thinking digital transformation executive with international and multi-industry experience, specializing in process excellence, business automation, enterprise architecture, intelligent and robotic process automation, continuous improvement, and business consulting. He is an expert in establishing and scaling practices, centers of excellence, and teams to drive simplicity, standardization, and optimization toward long-term growth. He is a savvy and resourceful leader who strives to facilitate the maturity of business transformation capabilities.

Partha is a strong believer that in addition to optimizing the core systems of record and systems of engagement, the transformation journey and goals can be accelerated by taking full advantage of emerging technologies such as Process/Task Mining, RPA/IPA (Robotic Process Automation/Intelligent Process Automation), Workflow Orchestration, Natural Language Processing (NLP), Document Processing (OCR), and Chatbots and by enabling/empowering employees to learn/adopt/leverage the power of these tools to be ready for the future of work.

Partha brings the right balance of strategic vision, team development, leadership, and execution focus along with deep experience in the business domains, processes, tools and systems, and data management to be effective.

In his leadership role in companies like VMware, UiPath, and HP, Partha implemented a comprehensive approach to deliver tangible business outcomes by establishing a scaled enterprise-wide Intelligent Automation Center of Excellence (CoE) and executing end-to-end process optimization and digitization, delivering intelligent automation solutions across all enterprise functions. He is also very passionate about "automation democratization" and has built one of the best-in-class citizen development programs to advance digital fluency and orchestrate process digitization and productivity optimization and low/no code automation adoption.

Bobby Jutley is the Interim Head of Digital Process and Automation at HP, where he leads an enterprise program to scale intelligent automation along with supporting establishment of a Process Transformation CoE.

Bobby has been on the automation journey since 2016 and has reinvented himself, going from a career in Finance to moving into Innovation and Digital Transformation. Bobby was instrumental in bringing UiPath to HP and adding the Microsoft Power platform and Process Mining technologies to the intelligent automation HP toolkit. Bobby has been a true champion for AI and automation and helped to define and steer the agenda at HP ground-up. Bobby is a 22-year veteran at HP and has worked across all facets bringing his rich experience to transformation efforts and initiatives corporate-wide. He has managed to balance corporate with academia and worked on research papers as he understands the importance of how automation will shape the future of work for generations to come.

Bobby is based in Houston and is heavily involved in community events with his passion for music. He participates in multiple forums and round tables on the topic of intelligent automation and AI and is excited about what the future holds as we embark on the power of AI and the speed at which technology is evolving.

Kriti Kapoor is a highly experienced executive in digital transformation with a global background in leadership roles across APAC, EMEA, and the Americas. She champions data-driven thinking and inspires cross-functional teams to deliver successful outcomes. At HP, Kriti leads worldwide Digital Process & Automation Community and Digital Championship initiatives, driving strategies for modernizing, digitizing, and automating operations while cultivating critical "future of work" skills. Her expertise in agile practices, business process management, and intelligent automation helps organizations become future-ready. Kriti is dedicated to building and scaling technical communities of key digital practices, promoting knowledge sharing and innovation across industries and geographies. She is also committed to advancing digital equity and diversity in the technology ecosystem. Kriti holds an MBA in Entrepreneurial Management and Marketing from London Business School and a BSc degree in Computer Science and Information Systems from the National University of Singapore.

Shail Khiyara is an accomplished leader in the field of Intelligent Automation (IA) and Artificial Intelligence (AI). He is a sought-after speaker, operator, and thought leader who has served on the executive team as Chief Marketing Officer (CMO) and Chief Customer Officer (CCO) at Automation Anywhere, Blue Prism, and UiPath. He is widely regarded as a unicorn in the automation industry and is known for his ability to drive growth and turnarounds amid acquisitions.

With a broad global experience in North America, Europe, and Asia, Shail brings a unique combination of global CEO-level operational experience with a blend of strategy, sales, marketing, customer success, and product domain expertise. He has more than two decades of enterprise software experience and has been instrumental in shaping the IA, RPA, and AI markets.

One of Shail's significant contributions to the automation industry is his founding of VOCAL, an independent, global, automation customer-only think tank. VOCAL stands for Voice of the Customer in the Automation Landscape and comprises the world's leading companies. VOCAL's aim is to provide a forum for automation customers to exchange best practices, share knowledge, collaborate, and advance the use of automation and AI technologies.

Apart from his work at VOCAL, Shail is also the President and Chief Operating Officer of Turbotic, a pioneer in Automation Optimization. He serves on the board of several automation and AI companies.

Dr. Tiffany Maldonado is an Assistant Professor in the Department of Management, Marketing and MIS, College of Business Administration at the Sam Houston State University. She earned her undergraduate degree in Mechanical Engineering and Managerial Studies from Rice University and her PhD from the University of Houston.

Prior to becoming a university faculty member, Dr. Maldonado worked professionally in a number of diverse industries such as Alagasco as an oil and gas engineer in the compliance department and Klien ISD as a fourth grade math and science teacher. These positions allowed her to gain experiences in the following areas: project management, oil and gas compliance, crowd control, motivating the unmotivated, inspiring hope, and changing fixed mindsets.

During her tenure in corporate, she became an expert in a software program called MAGI to keep inventory via GIS of gas piping and HCAs which are specially coded gas pipes that have a high consequence area. She developed and coordinated a student team who got first place in the Mars Rover Project.

Dr. Maldonado's passion to stretch the minds of others led her to later teach at the University of Houston-Downtown and Sam Houston State University in the following areas: project management, principles of business, performance management, talent development, leadership, negotiation, international business, and strategic management. She has been featured as a speaker at various conferences and leadership programs such as Course Hero, the PhD Project, and Babson's New Voices Program.

Dr. Maldonado's research streams of strategic leadership, organizational culture, and positive organizational scholarship provide interesting research questions on their own and combine to produce a fascinating research agenda. Dr. Maldonado publishes manuscripts in journals such as *Journal of Business Strategies*, *Business Horizons*, *Organizational Dynamics*, *Journal of Information Technology Teaching Cases*, *International Journal of Data Science*, and *IEEE Transactions on Engineering Management*. Her work has also been featured in *Salon* and the *Greater Good Magazine*.

Neeraj Mathur is the Director of Intelligent Automation at VMware, where he leads efforts to scale intelligent automation practices across the organization.

Neeraj is an AI and Automation Evangelist and has worked as the technical and strategic advisor to CXOs, senior executives, and center of excellence teams, enabling some of the world's largest organizations to scale through their intelligent automation and AI/ML journey. He is a technology leader who successfully built and managed scalable mission-critical equities trading, sales trading, and prime services systems with high availability and performance during his 15+ years at Goldman Sachs. Neeraj is an active member of the VOCAL Council and an Advisory Board Member of Intelligent Automation Select (IA Select).

Neeraj is based in Cupertino, CA. He serves as President of the Cupertino Historical Society & Museum to share local history and educational resources with students, scholars, and the public. Neeraj is passionate about contributing to the efforts to increase students' learning opportunities and enhance their academic achievements. He contributed through varied roles, from hosting a computer programming club sponsored by Google; mentoring students about college admissions, higher education, and career development; being the head coach of the local middle school's robotics team; and regularly judging for Speech and Debate tournaments.

Neeraj is pursuing a Master's degree in Information Technology Management from Carnegie Mellon University. He holds a Bachelor of Engineering (BE) in Computer Engineering from the University of Rajasthan, India. He has kept abreast through the Artificial Intelligence Graduate Certificate from Stanford University and executive certificates of Chief Information Officer and Chief Data Officer from Carnegie Mellon University.

Introduction

SINCE PREHISTORIC TIMES, HUMANS have invented ways to simplify daily activities to enhance our experience and improve productivity. The inventions well documented during ancient Greece or Egypt were targeted to automate for greater efficiency. The first significant step in the evolution of automation was to mechanize production through water and steam-powered automation, leading to the first industrial revolution. The second industrial revolution occurred when factories underwent electrification and used electric power to automate mass production. During this period, the manufacturing plants gained significant production increases as electric motors had much greater efficiency than steam engines. The advent of personal computers enabled the rapid growth of digital controls performing automation of more complex tasks at faster speed and greater efficiency, resulting in the third industrial revolution. In the current era of the fourth industrial revolution, the proliferation of digital devices and software systems has made automation relevant to the nonphysical work of the Information Age.

Evidently, the evolution of automation has always been driven by the quest to increase productivity, reduce costs, and increase efficiency. The most recent milestone of this journey is robotic process automation (RPA). The RPA helps to build software robots, also known as bots, that emulate human actions of typing using the computer keyboard, navigating software systems screens, recognizing and extracting the data on the screen, etc. These bots can be leveraged to automate mundane and repetitive tasks that can be labor-intensive and prone to errors.

In the past decade, digital native businesses have disrupted how corporations interact with their customers, partners, and employees. Data-intensive enterprises have harnessed the benefits of growing artificial intelligence (AI) and machine learning (ML) capabilities in their business processes for strategic and competitive advantage. Even traditional businesses have implemented ways to reinvent themselves to remain relevant and competitive by utilizing the power of AI/ML. This unprecedented rise of AI brings a new era of business transformation whereby business processes are enhanced and fueled through AI enabling data-backed decision-making across a range of small low-complexity tasks to highly complex cross-functional processes.

In simple words, integrating automation technologies (like RPA) with AI capabilities emulating human actions and decision abilities is referred to as intelligent automation (IA), or sometimes also referred to as cognitive automation or hyperautomation. Intelligent automation is a rapidly growing discipline that enables the advancement of business process automation by introducing AI into business workflows. The central concept of true IA is in combining and integrating existing and emerging IA technologies. Such integrations enable the full potential of IA and create a significant difference in making existing business processes efficient, accurate, and sustainable. Integrating technologies such as RPA, Process Mining, Advanced Analytics, Generative AI, Internet of Things (IoT), and Blockchain can play a vital role in augmenting business processes with AI-enabled capabilities facilitating data-driven decisions throughout the business process workflow and expanding the scope of process automation to cross-functional end-to-end process workflow.

This book provides practitioners and academics with the concepts, methods, best practices, and challenges to successfully plan, implement, and scale an IA program. This book is an unprecedented attempt to combine academic theories from leading academic researchers with practical insights from practitioners at the forefront of large and mature IA practices. With this unique collaboration of practitioners and academicians, this book attempts to bridge the gap between corporate and academia to help develop the talent pool of the future.

In this book, the authors address the critical questions about the rise, usage, and future of IA practices. How has the IA industry developed and risen to global success? How can IA technologies be accessed and leveraged? Why an organization's culture is critical for the adoption of IA? What actions can transform the organization's culture for successful adoption? Why is it paramount first to discover and standardize business processes? How to engage stakeholders using innovative and modern techniques for long-term engagement? What actions and approaches to consider in planning, establishing, and scaling IA practices? How to develop an outlook to manage the IA program effectively aligned with business objectives and priorities? What common pitfalls to avoid in building a solid foundation to scale IA practice enterprise-wide? How does the future of IA look from the authors' view? What avenues are available for practitioners and students to participate and contribute to the growth of IA? And many more.

WHO IS THE TARGET AUDIENCE?

This book is strongly recommended for practitioners seeking to plan, implement, and scale IA practices in their organization and for researchers and students who intend to study strategy, implementation, and management of IA practice to accelerate the digital transformation agenda.

To address the need of all readers, the authors have categorized the target audience into the following personas considering individuals' learning interests and areas of focus. In this context, these personas represent people with similar backgrounds or expertise in functional or technical fields and demonstrate a similar need to learn nuances of IA practice management. Though reading the book from cover to cover is strongly recommended, the authors have also emphasized a list of chapters that will provide the most relevant information immediately applicable to these personas.

Persona 1: Early-Stage Practitioners Seeking to Learn Effective Management of IA Programs

The individuals who have either embarked or are ready to embark on the journey of IA adoption seek to understand best practices for establishing IA programs to avoid common pitfalls. Their learning focus will be on the evolution of IA, its applicability, cultural influence and implication, business process standardization, implementation of best practices, and remediation of challenges. This book provides the methods and approaches to successfully apply cultural transformation, business process discovery, stakeholder management, implementation of best practices, and remediation of challenges.

Persona 2: Established IA Practitioners Seeking to Drive Maturity and Scale

This book is for individuals already partaking and managing IA CoE and who desire to learn more about the efficient management of the IA program to drive maturity and scale enterprise-wide. This book provides concepts, methods, and approaches for defining and delivering strategic value, monitoring strategic value delivery, establishing and maintaining strategic and architectural alignment, setting up organizational scale and operational scale, operationalizing support and maintenance, establishing and managing scaling dimensions enterprise-wide, and launching enterprise-wide citizen-led enablement.

Persona 3: Business Leader Seeking to Understand How to Drive Business Value Using IA

This book is for business leaders, no matter what level or function, who are interested in understanding how IA can add strategic and competitive value to their businesses. In addition, it supports and provides an understanding of the approaches for defining and delivering strategic value; monitoring strategic value delivery; understanding the importance of process standardization; establishing techniques for process discovery; determining the influence and implication of cultural alignment and how to achieve it; understanding the significance of CoE, the implementation of best practices, and how sponsors and stakeholders can help drive and support CoE initiatives for establishing, operationalizing, and scaling the IA program.

Persona 4: Practitioners or Academicians Seeking to Collaborate

This book is a result of a unique collaboration between corporate and academia. For individuals interested in developing or partaking in similar collaboration opportunities, this book presents a framework focusing on avenues, roles, and initiatives for collaboration. For example, these opportunities can help practitioners participate in educational programs to influence the future talent pool pipeline and identify opportunities to contribute to current research about digital transformation and management of IA programs. Similarly, these opportunities can help researchers or students participate in corporate initiatives to understand real-world use cases better, their functional applicability and challenges, and the limitations of IA technologies.

HOW TO READ THIS BOOK?

HOW TO READ THIS BOOK

CHAPTER OVERVIEW

1 **Chapter 1: The RPA Big Bang**
This chapter covers the history and rise of RPA/IA and how it continued to add business value.

2 **Chapter 2: The Value of Business Process Discovery**
To set up a successful IA program, you need to evaluate and transform your organization's business processes; learn more in this chapter.

3 **Chapter 3: Future of Work: Automation & Implications for Academia & Organizations**
To better understand the opportunities and approaches available for both corporate and academia to collaborate.

4 **Chapter 4: IA Accessibility and Applicability**
For a better understanding of how IA technologies can be leveraged.

5 **Chapter 5: Technology Only Does Not Yield Success – Humble Culture drives success in IA**
To set up a successful program, you need to evaluate and transform your organization's culture; learn more in this chapter

6 **Chapter 6: Intelligent Automation Implementation in Organizations**
This chapter will help you understand the IA program's scope, implementation, and best practices.

7 **Chapter 7: Challenges Scaling IA – The Race to IA Takes Off**
This chapter helps you to avoid common pitfalls by understanding common dimensions of scalability, the outlook to develop, challenges & solutions to scale.

8 **Chapter 8: IA Unleashed for Diversity – Botathons Become New Ways of Working**
Engage your stakeholders. This chapter helps unleash it through Botathons.

9 **Chapters 9: Automation is here to stay**
These chapters will provide you insight into the long-term applicability of IA programs and technologies.

10 **Chapters 10: Periscope into the Future**
These chapters will provide you insight into the future trends of IA assisting you to establish a long-term vision for your IA program.

A NOTE ABOUT PERSONAS

Personas represent people with similar backgrounds or expertise in functional or technical fields and demonstrate a similar need to learn nuances of IA practice management.

Persona 1
Early stage practitioners seeking to learn effective management of IA programs

Persona 2
Established IA practitioners seeking to drive maturity and scale

Persona 3
Business leader seeking to understand how to drive business value using full potential of IA

Persona 4
Practitioners or Academicians seeking to collaborate.

RECOMMENDED READING

Though reading the book from cover to cover is strongly recommended, the authors have also emphasized a list of chapters that will provide the most relevant information immediately applicable to these personas.

| | CHAPTERS | | | | | | | | | |
	1	2	3	4	5	6	7	8	9	10
Persona 1		X		X	X	X				
Persona 2						X	X	X		
Persona 3	X	X		X	X			X	X	X
Persona 4			X							

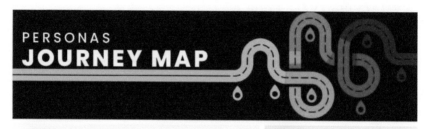

PERSONAS
JOURNEY MAP

 **PERSONA 1**

INTEREST: I'm new to Intelligent Automation / Digital Transformation and in process to initiate automation CoE and want to learn how to manage the IA program.

FOCUS: To learn about automation and digital transformation (history, applicability, process discovery, cultural implication, stakeholder engagement, implementation, challenges, etc.)

EMPHASIS: Chapter 2,4, 5 and 6 for business process discovery & standardization, cultural implication and transformation, implementation of best practices, lessons learned, challenges & solutions.

 **PERSONA 2**

INTEREST: We have established Automation CoE and would like to learn more about efficient management to drive maturity and scale.

FOCUS: To drive maturity and scale of the program (strategy & architecture alignment, value delivery, Operation & support, scaling dimensions & approaches, organizational scale, operational scale, citizen-led enablement, etc.)

EMPHASIS: Chapter 6,7, and 8 for implementation, lessons learned, and citizen development

 **PERSONA 3**

INTEREST: I'm head of a business function, business unit or an organization interested in understanding how IA can add strategic and competitive value for our business.

FOCUS: To sponsor or champion IA program (history, applicability, process discovery, cultural implication, stakeholder engagement, implementation, citizen-led enablement, etc.)

EMPHASIS: Chapter 1, 2, 4, 5, 8, 9, 10 for process standardization, cultural alignment, implementation and future roadmap.

 **PERSONA 4**

INTEREST: I'm a practitioner and would like to collaborate with academia to understand and contribute to current research. I'm a researcher or student interested in understanding and contributing to industry use cases.

FOCUS: Collaboration opportunities to participate in corporate initiatives, real-world use cases, and understanding the limitations of IA technologies. Participate in educational programs, influence future talent pool pipeline, and opportunities to contribute to current research and studies.

EMPHASIS: Chapter 3 for collaboration of corporate and academia

Key Terms and Definitions

As you read this book, you will find that specific key terms, described below, recur throughout the book.

Artificial Intelligence (AI) is the foundation for mimicking human intelligence by applying software algorithms built into a computing environment.

Business Continuity Planning (BCP)/Disaster Recovery (DR): Business continuity is defined as an ability of an organization to continue to deliver its products and services at predefined acceptable levels following a disruptive incident. The planning, testing, and actions to maintain business continuity are captured under business continuity planning. Similarly, disaster recovery is the process of recovering or reestablishing the ability to deliver after a disaster (natural or human-induced).

Business Process Discovery (BPD) is a way to understand business process mechanics and how the insights from this knowledge can be leveraged to reveal opportunities for applying automation and making business processes more efficient.

Business Process Management is the practice of modeling, documenting, analyzing, and optimizing business processes to align with strategic business objectives and priorities.

Business SME (Subject Matter Expert) is someone who provides critical knowledge and expertise in a specific business area.

Center of Excellence (CoE) is a dedicated team within an organization that focuses on developing and implementing technologies. This book uses this term to refer team dedicated to IA technologies.

Chatbot is a software product used to perform an online chat conversation using text or text-to-speech instead of having direct contact with a human.

Citizen Developer is a person who creates applications or automation meant to be consumed by themselves or their team members. They generally use low-code/no-code platforms to develop such applications or automation.

Computer Vision is a collection of techniques, such as object detection and image classification, enabling machines to interpret and understand visual information from images and videos.

Conversational AI is a set of AI technologies that enables chat messaging and speech-enabled human-like interactions with computer software applications.

Digital Transformation is a journey to adopt digital technology to digitize non-digital products, services, or operations.

Digital Workers/Digital Workforce/Bots are digital (software) workers that augment human workers by combining RPA and AI/ML capabilities to optimize and automate processes.

End-to-End Process refers to the complete business process from start to finish, which might span over multiple teams or multiple business functions.

Hackathon/Bot-a-Thon is an event where people get together for a short period (1–2 days or less) and engage in collaborative design and development of bots to showcase their knowledge and learning while developing something meaningful.

Intelligent Automation: Integrating automation technologies (like RPA) with AI capabilities emulating human actions and decision abilities is referred to as intelligent automation (IA), or sometimes also referred to as cognitive automation or hyperautomation.

Intelligent Automation Program/Practice is meant to drive the adoption of IA technologies to achieve business objectives and goals. This practice follows methods and approaches to drive the maturity and scale of IA adoption enterprise-wide while delivering business value aligned with business priorities.

Intelligent Document Processing (IDP) platforms can convert unstructured and semi-structured data from documents into structured and usable information.

Key Performance Indicators (KPIs) are quantifiable measures of performance meant to evaluate, measure, and monitor the success of an organization or a particular initiative or activity.

Low-Code/No-Code development platforms work on the principle of reducing software development complexity by leveraging visual building blocks such as drag and drop. In addition, these platforms use techniques like code templates, plug-ins, and graphical connectors to automate a significant portion of the development process.

Machine Learning is a subfield of AI, broadly defined as the capability of machines to imitate intelligent human behavior.

Natural Language Generation (NLG) is a field of AI that produces written or spoken natural language narratives.

Natural Language Processing (NLP) is a field of AI that allows machines to read, understand, and derive meaning from human languages.

Optical Character Recognition (OCR) software systems are meant to read and convert printed, typed, or handwritten text into machine-readable digital format.

Process/Business Process is a collection of activities or tasks performed by employees of an organization to provide services related to the business function these employees are employed in.

Process Mining is a technique designed to discover, monitor, and improve actual processes by extracting readily available knowledge from the event logs of the information system. Process Mining enables organizations to gain a deep insight into their business processes and can be critical in simplifying and optimizing processes.

Proof of Concept (PoC) is an exercise performed to demonstrate the feasibility of a specific idea, method, or product.

Proof of Value (PoV) is a PoC that emphasizes the expected business value of a specific solution.

Return on Investment (ROI) is a performance measure used to evaluate the efficiency of an investment.

Robotic Process Automation (RPA) helps to build software robots, also known as bots, that emulate human actions of typing using the computer keyboard, navigating software systems screens, recognizing and extracting the data on the screen, etc.

Service-level Agreement (SLA) is a contract that defines and documents the level of service a customer expects from a service provider.

Task Mining is a technique designed to collect and explore user data in real-time interactions with the process. Task mining collects the user's clicks, keystrokes, and other interactions with the front end of the business application.

CHAPTER 1

Introduction

The RPA Big Bang

Shail Khiyara

Turbotic

A UTOMATION HAS A LONG and storied history dating back more than 1,000 years. The evolution of intelligent automation (IA) has been driven by the rapid advancement of technologies, and organizations worldwide are building their digital workforce. The countdown has begun, and depending on which report, article, expert, or practitioner you interact with, the projected growth numbers of digital workers are up and to the right.

Automation is about humans and bots working together to create a hyper-productive digital workforce, and the human potential is enormous.

Waves of automation have occurred in the past. The first era was in the 19th century when machines took away dirty and dangerous tasks leading to the creation of industrial equipment from looms to cotton gins relieving humans of manual labor. In the second era, around the 20th century, machines took away the dull, with automated interfaces and airline kiosks to call centers, relieving humans of routine service transactions. In this era, the third one, machines take away decisions with intelligent systems from airfare pricing to artificial intelligence (AI) solutions that could make better, faster choices than humans (Khiyara, 2018).[1]

Computerized automation in the 1970s led to the creation of management information systems (MIS) groups within organizations, including continuous quality improvement methods such as total quality management (TQM). In the 1990s, business process management (BPM) arose to optimize end-to-end (E2E) processes leveraging labor arbitrage, and as tech companies sought entry into this market, they brought along with it technology to enable Business Process Automation (BPA).

While BPA continues to grow, it is projected to be a $20B market by 2026. The growth of BPA led to the birth of RPA and, in fact, when Blue Prism started focusing on the BPO/BPM space as a path to market before they coined the term Robotic Process Automation (RPA). While business process outsourcing (BPO) had as one of its roots a labor arbitrage advantage, RPA soon sought to reduce costs in BPO and quickly started to gain adoption within shared services.

While automation dates back centuries, RPA as a term was coined in 2012. The growth of RPA has been driven by the increasing demand for efficient, cost-effective, and scalable solutions for automating repetitive and routine business processes. RPA involves the use of software robots, or "bots," that can be trained to mimic the actions of human users, such as clicking, typing, and navigating applications, in order to automate a wide range of business processes.

The market saw modest growth initially, and the first Forrester Wave app in RPA came out in 2016. Use cases revolved around the banking and financial services market back then, and this sector still carries the highest adoption rate. RPA as a sector saw a large infusion of cash in a short period of time. Between 2017 and 2020, roughly $11 billion of investment went into this market, including a set of ~15 acquisitions in the space of 2 years (2019–2021) for a total acquisition cost of approximately $1.5B. By 2021, there were ~80 RPA companies in the market. This was followed by massive, rapid hiring, shrinking, less than attractive unit economics, and over 50% of automation programs not seeing signs of success. The primary reason is that while the technology offers value, it is not as easy to deploy and needs expert talent. Automation universities and free training emerged as offerings from the incumbent RPA players.

As use cases expanded to include computer vision to read documents, intelligent document processing (IDP) emerged as a sub-sector of automation. IDP uses AI and machine learning (ML) algorithms to extract information from structured and unstructured documents such as invoices and contracts. RPA was the hands and legs, and now, automation had eyes through ML capabilities.

Another technology that emerged in the 2017 timeframe as an adjacency to RPA, though it has roots dating back many decades, is process mining. Process mining uses data mining techniques to analyze the logs of business processes, such as the flow of documents or the steps involved in a particular task. This allows organizations to identify bottlenecks and inefficiencies in their processes and optimize them for better performance. Often termed as the enabler for RPA with its ability to identify processes to automate, the process mining market saw multiple entrants in the market, leading to a faster consolidation in the market compared to what we say in RPA.

Over time, RPA has evolved and been combined with other technologies, such as AI and ML, to create what is now known as "intelligent automation." This type of automation is more advanced than traditional RPA, as it uses AI and ML algorithms to make decisions and perform actions. Intelligent automation can handle more complex tasks and make decisions based on a wider range of data. The development of IA was driven by several factors, including the increasing availability of data, the development of powerful computing technologies, and the need for businesses to improve efficiency and reduce costs.

Hyperautomation, yet another new term, was coined by Gartner in 2020 as the number one technology trend. Hyperautomation uses AI and ML to increasingly automate processes and augment humans, not too different from IA. Gartner further classifies it as a business-driven, disciplined approach that organizations use to identify rapidly, vet, and automate as many business and IT processes as possible.

Amid this labeling journey from RPA to IA to hyperautomation, the Centre of Excellence (CoE) has played a major role in driving the success of automation for organizations. The growth of the CoE in automation has been a trend that has been gaining traction in recent years. A CoE is a dedicated team or unit within an organization that focuses on developing and implementing technologies. These teams often have a mandate to drive the adoption of automation across the organization and to help other teams leverage automation. Several different models of CoEs have emerged, with the most common being a centralized CoE which allows for automation to be controlled by the central CoE and can be beneficial in terms of coordination and consistency.

Another model is the decentralized CoE, in which multiple teams within the organization are responsible for developing and implementing automation technologies. Hybrid models, though rare, are also seen in practice.

This explosion of initial adoption, adjacent technologies, and increased valuations has also been met with several barriers to automation. Resistance to change is by far the biggest that organizations struggle with even today. In its quest for rapid growth to match the investment outcome expectations of this market, two critical factors have been overlooked.

One has been around driving customer lifetime value. The risk of automation success is entirely borne by the customer, who must leverage system integrators to implement automation and help make it a success. Several have done it on their own with moderate, outlier success. The other is around a standardized way to measure automation success. What started with full-time equivalent (FTE) reduction as a key metric then morphed into hours returned back to the organization to multiple other derivative forms of metrics. For example, revenue went up because customer satisfaction went up due to the customer success person having more time to drive empathy in the conversation because they had something automated on their desktop. I published 60 plus automation metrics, a series that might be helpful.

As of the publishing of this book, organizations will have grown to a multi-vendor footprint and a collection of technologies to drive automation. Amid this journey of the current era in automation, there are excellent market insights, customer stories, and, most importantly, the opportunity to prepare our current and future workforce with shared challenges and lessons learned.

Equally important is the ability to skill our future workforce in the science and art of automation, bridging the gap between business partners and academia. Digital transformation absolutely requires and involves people transformation. This book is designed to help you transform your organization and culture with your most important asset—your people at the core.

NOTE

1 https://medium.com/@ShailKhiyara/countdown-to-a-digital-workforce-2060b4566712

CHAPTER 2

The Value of Business Process Discovery

Bobby Jutley

HP Inc.

INTRODUCTION

This chapter talks about the importance of business process discovery (BPD) as a discipline to help drive value realization for your organization. It introduces the concepts of process mining and task capture and discusses the approaches and benefits of including them in your business process discovery armory. Using these tools helps you better understand your processes' current health and ultimately helps identify opportunities for process improvement.

As more and more companies encounter challenges from the convergence of social, mobile, cloud, and smart technologies, they have put digital transformation at the top of their priority lists as they strive to drive more innovation, automation, and integration. As a result, the digital strategy will be at the core of everything, and this needs to be complemented by strong digital expertise, agility, and collaboration across the organization.

Business process discovery really helps to create an open-minded culture that facilitates creativity and risk-taking, and this is what is key to driving a successful digital transformation program.

DOI: 10.1201/9781003276128-2

WHAT IS BUSINESS PROCESS DISCOVERY?

Business process discovery has become an increasingly popular way to understand your process mechanics and how the insights from this knowledge can be leveraged to reveal opportunities for applying automation and making your *Process* more efficient. This notion of business process management and process mining relies on a set of techniques representing an organization's current business processes. This also highlights potential process variances and provides an opportunity to standardize such variances.

Today, businesses need to better understand process management and the opportunities for process improvement. This is necessary to see where improvements can be made to the (to-be) *Process* and gain a greater understanding of the as-is construct. By doing this, they can, first of all, decide whether improvements can be made and, more importantly, whether it is worth investing in these improvements.

The process discovery techniques capture data recorded as the *Process* is executed in its normal manner leveraging the existing organizational methods of work, documentation, and technology systems that run the business processes within an organization. These data help to understand better the underlying process model, resources executing the *Process*, the different types and nature of events, and essentially an end-to-end granular view of the *Process*. Business organizations often understand a process but don't have the tools or cannot accurately articulate possible deviations and bottlenecks in a process, which ultimately negatively impacts many fronts.

A general problem with process management is the inability to connect business and IT systems. Many enterprise systems are fine with supporting transactional processes, for example, order-to-cash or procure-to-pay, but connecting the actual *Process* with the information system is not straightforward. Collecting information about how your *Process* is performing on a day-to-day basis requires a complex set of manual steps to gather and synthesize data. Process modeling, mapping, and determining the most effective and efficient *Process* require detailed analysis and modeling. This is where we have seen the discipline of process mining evolving (Davenport & Spanyi, 2021).

Management will have a certain expectation of a *Process*, but unfortunately, the reality is somewhat different. Hence, before any potential automation can be applied to a process, it is extremely important to understand all the steps that are taken for your process. Figure 2.1 depicts process expectation vs. process reality.

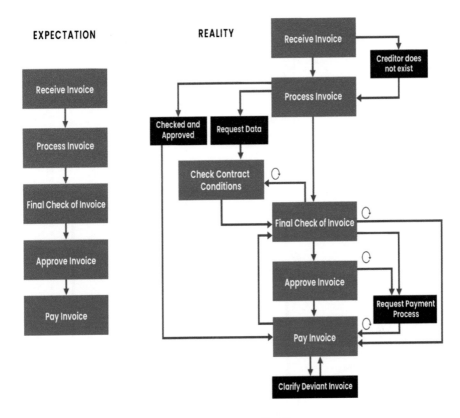

FIGURE 2.1   Process Expectation vs. Process Reality.

WHAT IS PROCESS MINING?

The process mining discipline emerged in the late 1990s at the Eindhoven University of Technology with the pioneering work of *Prof.dr.ir. Wil van der Aalst.* He is widely known as "the Godfather of Process Mining" and has influenced the maturing of the discipline. Today, over 35 commercial process mining vendors and thousands of organizations are successfully applying process mining (Homepage Wil Van Der Aalst, n.d.).

Process mining can be thought of as the bridge between process science and data science. The software ultimately helps to turn the data into insights and actions. For example, what is the *Process* that people follow, do people really follow the expected *Process*, and are there any bottlenecks in my *Process*? These are all questions that the insights from the software can help answer. The software can help organizations easily capture information from enterprise transaction systems and provide detailed—and data-driven—information about how key processes perform (see Figure 2.2 for more detail).

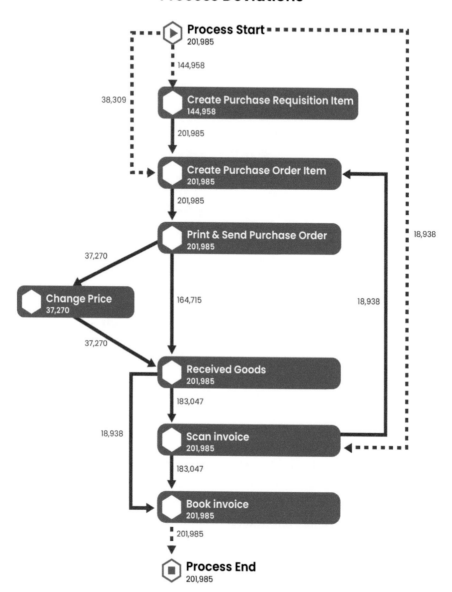

FIGURE 2.2   Process Mining Digital Map.

It creates event logs as work is done: an order is received, a product is delivered, and payment is made. The logs depict how the work is actually happening, for example, who carried out the activity, how long it took, and how it departed from the baseline. The process analytics can then be performed using pre-configured algorithms and models that provide key performance indicators for the process. This, in turn, enables a company to focus on where there is an opportunity for improvement and prioritization around going after these opportunities (Davenport, 2019).

Process mining can be categorized into four stages:

1. **Data collection**: source system data retrieval/event logs/data visualization using real-time system integration

2. **Data discovery**: analysis of the data that have been captured

3. **Process enhancement**: automate tasks based on inputs from the data discovery stage

4. **Process monitoring**: continuously track and monitor key metrics ensuring performance and process conformance

As organizations have matured in their automation journeys, the pipeline for identifying robotic process automation (RPA) opportunities has become more challenging, not to mention how you should prioritize your solution deployments. One solution becoming increasingly popular in understanding process complexity is task capture discovery (Kosmopoulos, n.d.).

WHAT IS TASK CAPTURE?

Task capture uses technology to help better explore, understand, and analyze the tasks that employees carry out as part of a bigger end-to-end process during their day-to-day activities.

This is done by installing the task-capturing software on the employee's computer. In turn, the software monitors and captures the transactions and actions that the employee carries out across the different systems, landscapes, and overall ecosystem as part of their job. In addition, the software records all the interactions the employee does to perform their tasks, which are all automatically tracked. This recording includes data such as mouse clicks, data input, and keystrokes, and it can be controlled in terms of what applications the user wants or needs to have recorded.

The purpose of task capture is to gain a better understanding of what tasks the employees are performing. If you think of an employee, they all have their unique fingerprint in the way they carry out a task, and it is this uniqueness that task capture software analyzes. It identifies where improvements can be made to improve operational efficiency, error reduction, employee morale, and opportunities for automation.

WHAT'S THE DIFFERENCE BETWEEN TASK CAPTURE AND PROCESS MINING?

The difference between both capabilities is the focus on how processes and tasks differ. *Processes* are logical and typically have a sequence that makes up the process. For example, when we look at the accounts payable process, the processing of an invoice can be clearly defined as a process that is made up of constituent sub-processes, such as the approval of the payment. On the contrary, *tasks* are a collection of related steps within a process. In other words, a number of steps are required to execute a process. For instance, if we take the processing of an invoice, tasks related to that process would include extracting the data from the invoice, which requires multiple steps (see Figure 2.3).

Task capture and process mining should be used in tandem, where process mining discovers the organization's entire business processes, such as accounts payable and accounts receivable, and task capture records and finds how the individual steps that make up that *Process* are completed (Kosmopoulos, n.d.).

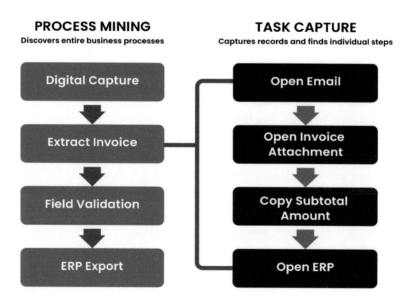

FIGURE 2.3   Process Mining vs. Task Capture.

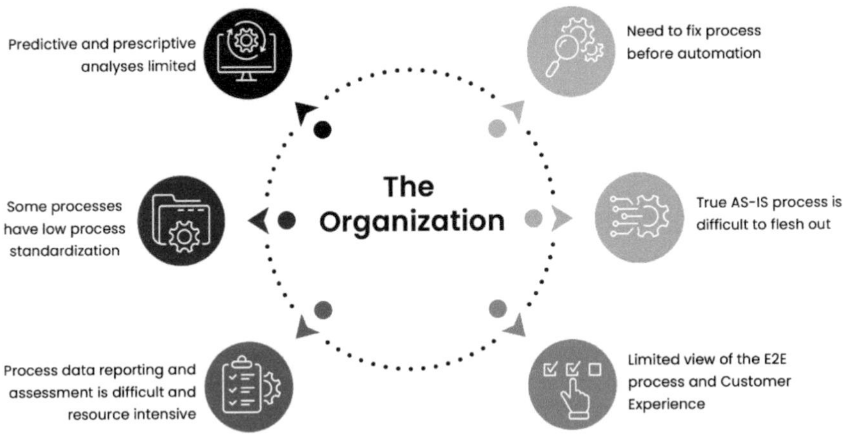

Predictive and prescriptive analyses limited

Need to fix process before automation

Some processes have low process standardization

The Organization

True AS–IS process is difficult to flesh out

Process data reporting and assessment is difficult and resource intensive

Limited view of the E2E process and Customer Experience

FIGURE 2.4   Benefits of Process Mining.

WHY SHOULD AN ORGANIZATION CARRY OUT PROCESS MINING?

For an organization to be successful, it needs a full and foundational analysis of all its processes and must bring in a holistic picture to be able to drive transformation. These process analysis insights provide an opportunity for re-engineering and improving the current process. By applying process mining, the data can provide more predictive analytics, highlight process deviation, drive automation, and help to connect an end-to-end view of the process, as depicted in Figure 2.4.

When we think about digital transformation, companies tend to embark on a digital transformation agenda without having a clear definition, let alone a vision for what it means (SGSubra, 2022). Chamorro-Premuzic calls out five essential components of a digital transformation (Davenport & Spanyi, 2021) (see Figure 2.5).

As shown in Figure 2.5, data are a key element for digital transformation success, and as mentioned before, if we want to digitize and automate our processes, data are a key component we require to gain a better understanding of the following:

1. Automation opportunities that are hidden within a process

2. A holistic approach to automation

3. The potential for converting insights into action

4. The potential for monitoring processes and keeping them on track

The 5 Essential Components of a Digital Transformation

Mapping the journey to becoming a data-centric organization.

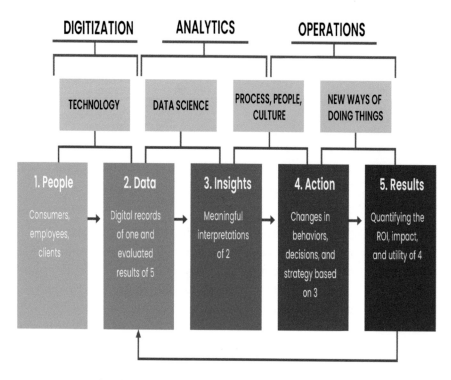

DIGITIZATION		ANALYTICS	OPERATIONS	

TECHNOLOGY	DATA SCIENCE	PROCESS, PEOPLE, CULTURE	NEW WAYS OF DOING THINGS

1. People	2. Data	3. Insights	4. Action	5. Results
Consumers, employees, clients	Digital records of one and evaluated results of 5	Meaningful interpretations of 2	Changes in behaviors, decisions, and strategy based on 3	Quantifying the ROI, impact, and utility of 4

FIGURE 2.5 The five Essential Components of a Digital Transformation. (Adapted from The Essential Components of Digital Transformation – HVR org.)

This is why the idea behind process mining is to really get to the "weeds" of the *Process* and understand what the data are telling us and what we should infer from the data being captured.

HOW DOES PROCESS MINING HELP IN YOUR AUTOMATION JOURNEY?

In 2020, Gartner (see Gartner Market Guide for Process Mining Nov 2021 by Marc Kerremans, Tushar Srivastava, Farhan Choudhary) estimated the process mining market for new product license and maintenance revenue to be $550 million, which was over 70% of market size growth from the previous year. The process mining market is forecast to grow between 40% and 50% and will surpass $1 billion in 2022 (Kanhonou, 2022).

In the same report, Gartner talks about the following key findings:

- Traditional interview-based process discovery and modeling are costly and time-consuming because of gaps in business knowledge, a lack of objective validation techniques, and poor formalisms.

- Formal standard operating procedures, policies, work instructions, or best practices baked into enterprise applications are often compromised by informal behavior that bypasses governance.

Gartner has identified 10 capabilities for process mining for:

1. **Models of processes, exceptions, and process instances (mostly referred to as "cases"), and employee interactions** — Automated Discovery of process models, exceptions, and process instances, together with basic frequencies and statistics

2. **Support for customer interactions, customer journey maps, and related analysis** — Automated Discovery and analysis of customer interactions, as well as alignment with internal processes

3. **Conformance-checking and gap analysis capabilities** — Capabilities to check conformance not only graphically through overlays but also through data analysis and performing gap analysis

4. **Intelligent support for process model enhancement** — Enhancing or extending existing or a priori process models by using additional data from the recorded logs and events

5. **Data preparation and data cleansing support, supporting big data** — Different ways to handle data

6. **Real-time dashboards with support for key performance indicators (KPIs) that are continuously monitored and enable decision support** — Real-time or near-real-time connections to continuously monitored and adapt KPIs in dashboards for specific roles in the organization

7. **Predictive analysis, prescriptive analysis, scenario testing, and simulation** — Advanced analysis capabilities that use contextual Data

8. **A platform that extends the process mining capabilities across different processes with advanced analytic capabilities and decision management capabilities and that also offers APIs to create process mining apps** — Allowing organizations and partners of the process mining vendors to create applications, such as financial auditing tools

9. **Task mining** — Inferring useful information from low-level event data available in UI logs. These UI logs describe the single steps within a task done by a user—for example, when using a workstation—based on keystrokes, mouse clicks, and data entries.

10. **Execution capabilities that turn "insights" into "action"** — These capabilities could range from simply updating source applications (applications that delivered the events for process mining) to creating scripts that support the execution of tasks.

Gartner, Market Guide for Process Mining, 11 November 2021, Marc Kerremans, Tushar Srivastava, Farhan Choudhary

The industry trends point to value outcomes from process mining, and this discipline should be carefully considered as part of your business process discovery and automation strategy.

When we look at a process, we are not completely sure of the quality or efficiency of the process or whether there is room for improvement. Process mining will help to determine the health of the process so that Process Mining can be thought of as the gateway to automation.

As you embark on your automation journey, process discovery is key in assessing the types of processes you should go after for automation. When evaluating processes, you will typically look for headcount-intensive processes that lend themselves to high automation potential rates, repetitive task-oriented processes, logical steps, and low-value activities—this will be covered in more detail later on in this book.

Business process discovery will benefit your organization in terms of the following:

- **Scalability**: As you map out your process, you can deploy RPA solutions faster.

- **Cost reduction**: The mapping and workflow outputs reduce development time and cost.

- **Quality improvement**: It provides better insights and continuous improvements.

- **Risk reduction**: Overall, it provides tighter control with fewer people having access to data.

In order to get the required value outcome from process discovery, a significant investment is required in terms of licensing cost, resource cost, and time and effort. This investment cannot be underestimated and needs to be carefully considered. Therefore, both technical teams and the business need to jointly agree on the strategy around process discovery and how they would like to apply process mining techniques.

One will have to consider the following questions as a starting point:

- How much should I invest in licensing?

- What processes should I go after to reap maximum returns from process discovery?

- Do I have the buy-in from the business to help support my process discovery efforts?

- Have I identified all the required stakeholders, and are they fully engaged? For example, IT will help support your data storage needs.

- What are the success criteria for return on investment (ROI)?

Once you have clarity around these questions, you should evaluate the market, consult with different vendors, and have them take you through their process to discover capabilities. Not all vendors include all capabilities, so you need to carefully consider what you are looking for and what product best fits your requirements.

You will then jointly decide what process makes sense to do a proof of concept (PoC) or a proof of value (PoV). As part of this PoC, you will have to clearly call out your success criteria, which should include specific use cases to be tackled and your potential ROI targets to be achieved.

EXAMPLES OF USE CASES FOR PROCESS MINING

Process mining use cases have been applied across many companies, and companies are realizing the benefits, whether related to contact service support, finance, supply chain, sales, and marketing. However, all these

TABLE 2.1   P2P Use Cases

Duplicate invoice/ payments	• Enable real time/proactive and efficient suspect duplicate check • Avoid duplicate payments
Pay term compliance	• Identification of non-standard payment terms without required approvals • Enabling reporting around non-standard payment terms and impact on working capital/DPO • Enabling capability to compare the vendor payment terms from contracts to payment level
On-time payment (OTP) compliance	• Enable capability to report and analyze the reasons for missing on-time payment
Withholding tax compliance	• Enabling reporting/analysis capabilities for withholding tax compliance • Avoid wrong/non-deductions resulting in rework and compliance exposure
Vendor master accuracy	• Enable real-time capability to analyze duplicate vendors • To be able to actively Validate the accuracy of vendor data across systems (master data systems & business instances)

functions produce business challenges that need to be root-caused to become more standardized and efficient to generate more value and a better customer experience. Table 2.1 provides examples of use cases for procure to pay (P2P) that help to drive value.

THINGS TO CONSIDER WHEN EMBARKING ON AN INTEGRATED PROCESS DISCOVERY JOURNEY

At the start of this chapter, we discussed the importance of business process discovery and how this fundamentally steers you into a mindset change of continuous improvement and value capture. Getting a deep understanding of your organization's processes is becoming more and more relevant to establishing any baselining from where changes can be made to improve efficiency, effectiveness, and execution. This all plays a part in improving the overall customer and employee experience, which is the secret sauce to a successful company.

The less efficient a *Process*, the more expensive it is to run and the longer it will take to realize value and impact revenue. As an organization, companies must make deliberate decisions around the importance associated with this effort. In addition, investments need to be made in systems, resources, and software, and the organization needs to align on KPIs and success metrics that it wants to impact and improve directly. Finally, the company needs to clearly define a strategy, allowing time for the team to embrace the mindset change, get established, and then focus on achieving the desired results.

As you review opportunities for process re-engineering and potential automation, you will need to:

1. Understand the customer journey and really understand how this aligns with the *Process* you are looking to improve

2. Understand the process variations and gain relevant insights and intelligence

3. Incorporate both offline and online process steps into the process mapping view to better understand where there will be opportunities to streamline and improve the *Process*

4. Perform conformance checking to compare and contrast process standardization deviation

5. Define opportunities for process improvement and automation

6. Carryout in-depth analysis for solutions

7. Plan and deliver the solutions, document the newly re-engineered process steps, and continue to monitor and sustain with active dashboard reporting

An example of an E2E blueprint approach from process discovery to value realization can be seen in Figure 2.6 and used as a framework as you drive your transformation.

SUMMING UP

Hopefully, this chapter has helped you understand the practical application and approach to business process discovery and its disciplined approach to driving value for the company. The discipline of process discovery requires a radical mindset change as companies try to balance business as usual priorities along with delivering digital transformation goals. This balancing act can cause conflict and frustration as employees and teams will always have varying degrees of process-maturity understanding and willingness to invest and put effort into establishing a heatmap and work through process transformation.

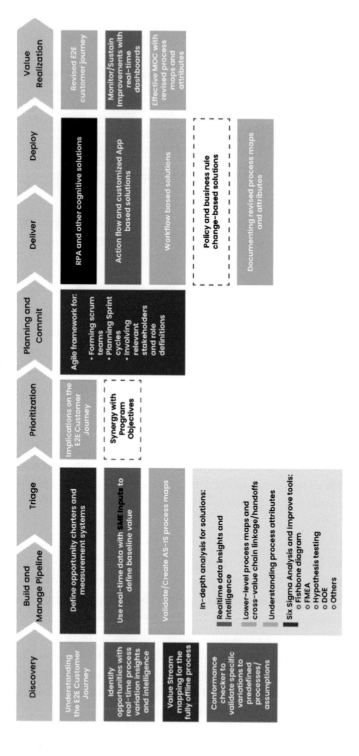

FIGURE 2.6   An Integrated E2E Process Discovery Approach to Drive Value Realization.

Process mining, task mining, and business process management all play an integral part in process discovery. You will have to find the right constituents to weave your strategy and overall priorities into a beautiful tapestry as you embark on your RPA journey. Equally important is that if you want to scale RPA across the company and achieve your savings targets, you will have to evolve your process discovery skills and capabilities.

To conclude, insights derived from business process discovery are critical to managing, controlling, and improving your business processes and value realization. As you look to re-engineer, automate, and digitize your *Processes*, with the help of technology, business process discovery is on the critical path to achieving greater success.

CHAPTER 3

Future of Work

Automation and Implications for Academia and Business

Carol Brace

Sam Houston State University

Lila Carden

University of Houston

Kriti Kapoor

HP Inc.

Tiffany Maldonado

Sam Houston State University

INTRODUCTION

Organizations are using digital transformation as a strategic imperative to accelerate creating business value during times of tremendous economic uncertainty. This necessitates organizations to develop greater digital relevance and efficiency as the use of automation intensifies. In addition, the division of labor is shifting between humans and machines. As a result, the need for new digital skills that support the digital transformation agenda is growing, and new digital roles within organizations are emerging. However, while the demand for digital skills is high, the supply is low, and businesses are unable to recruit and timely train individuals with essential

DOI: 10.1201/9781003276128-3

digital skills. Moreover, educational systems globally are not currently prepared to equip individuals for the future of work in a digital era. Therefore, the authors recommend a framework including a partnership between organizations and academic institutions to close the digital skills gap in a timely manner and prepare a competent digital workforce for the future.

DIGITAL TRANSFORMATION

Organizations are dramatically increasing investments in digital transformation by using digital technologies and new ways of working to enhance customer, partner, and employee experience, transforming functions such as supply chain, marketing, sales, operations, and human resources. Business executives are driving investments in advanced data analytics, business process optimization, reengineering and digitization, and intelligent automation for greater efficiencies and business resilience. In addition, with the continued rise in digital adoption, tremendous business model transformations are underway as organizations accelerate efforts to bring new digital products and services to market. Three key trends, specifically (a) an increase in digital intensity, (b) an increase in digital transformation, and (c) an increase in automation, present unique challenges and opportunities for business leaders and employees worldwide.

Increase in Digital Intensity

Corporate longevity and relevancy are declining, and the average lifespan of a company on the Standard and Poor's (S&P) 500 Index is now 21 years instead of 32 years in the 1960s (Statista, 2021). Predominantly, digital leaders and digital platform providers dominate today's S&P 500 by market capitalization, including Apple, Microsoft, Alphabet, Amazon, Tesla, Meta, and Nvidia in the United States of America and the likes of Tencent and Alibaba in China.

Digital intensity, including the level of investment in technology-enabled initiatives, has and will continue to rise, impacting companies and sectors globally and thereby accelerating the need for digital transformation. See Figure 3.1 S&P list of top companies by market utilization.

Increase in Digital Transformation

Organizations are dramatically increasing their investments in digital transformation, using digital technologies and new ways of working to re-engineer and create new business processes, transform their culture and operating environments, adapt to dynamic market conditions,

	1980	1990	2000	2010	2022
1	IBM	IBM	GE	Exxon	Apple
2	AT&T	Exxon	Exxon	Apple	Microsoft
3	Exxon	GE	Pfizer	Microsoft	Amazon
4	Standard Oil, Indiana	Philip Morris	Citigroup	GE	Tesla
5	Schlumberger	Shell	Cisco	Chevron	Alphabet (Google)
6	Shell	Bristol-Myers	Walmart	IBM	Nvidia
7	Mobil	Merck & Co	Microsoft	Procter & Gamble	Berkshire Hathaway
8	Standard Oil of Cal	Walmart	AIG	A&T	Meta (Facebook)
9	Atlantic Richfield	AT&T	Merck & Co	J&J	J&J
10	GE	Coca-Cola	Intel	JP Morgan Chase	United Health

FIGURE 3.1   S&P Top Ten by Market Capitalization (1980 to Current). Adapted from Hunkar D. (2021) and The Top 25 Stocks in the S&P 500 (2022).

and accelerate product and service innovation customer experiences. Furthermore, at least 70% of organizations have a digital transformation strategy in place or are working on one. As a result, the global digital transformation market is expected to grow from $470 billion in 2020 to $1 trillion by 2025 (Caindec, 2021). Digitally transformed organizations are projected to contribute $53.3 trillion, or more than half of the global gross domestic product (GDP), by 2023 (Statista, 2022). Additionally, research predicts that 65% of the world's GDP will be digitized by 2022 (IDC FutureScape: Worldwide Digital Transformation 2021 Predictions, 2021).

Increase in Automation

According to a recent World Economic Forum Future of Jobs Report, published in October 2020, 84% of companies are accelerating the digitalization of their work processes, and 50% of organizations are increasing the use of automation to keep pace with rising customer expectations, improve productivity, and transform quicker (The Future of Jobs Report 2020, n.d.)

Manual jobs and daily repetitive, tedious tasks are more susceptible to automation opportunities. Almost half of the knowledge work activities performed globally have the potential to be automated using current low-code/no-code tools and RPA technology. It is estimated that 60% of all occupations have at least 30% technically automatable activities, such as data capture, processing, reconciliation, validation, and report creation (World Economic Outlook Databases, 2020). More specifically, office administrators and business analysts perform many of these activities in operation roles, including sales, marketing, finance, human resources, supply chain, and customer support.

DIGITAL READINESS

According to industry analysts, digital transformation is taking organizations twice as long and costing at least twice as much, partly due to their workforce's lack of digital readiness. This includes employee fluency with new tools and new digital ways of working.

More recently, the 2023 World Economic Forum shows upcoming shifts in the job landscape over the next five years:

- Technology and digital access are top drivers of transformation for 85% of organizations.

- Big data, cloud computing, AI, and digital platforms are predicted to impact job creation positively.

- Green investments have the strongest net job-creation effect, while agriculture tech, digital platforms, e-commerce, and AI will disrupt labor markets.

- By 2027, 42% of business tasks are expected to be automated, resulting in 26 million fewer jobs in record-keeping and administrative roles.

- 44% of workers' skills will be disrupted in the next five years, and companies are investing in learning, training, and automation to achieve business goals. (*Future Jobs: These Are the Fastest Growing and Fastest Declining Roles*, 2023)

For companies to take advantage of the newly created jobs, their workforce must be prepared with new skills and capabilities. This digital revolution will require at least 50% of current employees to be re-skilled, and 40% of their core skills will require a change to meet rising digital intensity. As a result, businesses expect to face the daunting task of reskilling current employees. Currently, 76% of employees indicate they are not prepared for the digital economy (Solis, 2022). Furthermore, this reskilling needs to happen at all levels of the organization, as only 23% of CEOs are considered digitally fluent, and 7% of companies believe that their executive teams are digitally fluent (Companies with a Digitally Savvy Top Management Team Perform Better, n.d.).

Companies will need to develop resilient leaders who are equipped to coach and support employees through a humble and agile corporate culture. They need digitally fluent employees who are comfortable in a hybrid work environment and working with artificial intelligence. In the midst of the digital revolution, employees still need uniquely human skills such as creativity, emotional intelligence, leadership, and critical thinking (2021 Workplace Learning Report, 2021) to navigate the ever-changing, ambiguous business environment.

To prepare new employees for the job demands of the digital revolution, organizations and academic institutions need to significantly advance their educational curriculum, training, and development activities to close the skills gap of the future. Studies have shown that only 29% of new hires have the essential skills to perform their jobs. This does not take into account that due to the digital transformation, many of the current skills being taught will soon be considered obsolete (Reengineering the Recruitment Process, 2021). Therefore, business leaders must purposefully plan the future of their organizations as well as the readiness of the future workforce. Likewise, academic institutions and organizations must intentionally and rapidly shift today's curricula to meet the accelerating industry demand and prepare for future jobs (roles).

Not only will businesses lose out if these skill gaps are not met, but their countries will miss out on cumulative GDP growth. Moreover, 14 of the G20 countries could miss out on $11.5 trillion in cumulative GDP growth if the digital skills gap is not closed. This can negatively impact product development, delivery, innovation, customer experiences, and satisfaction (Salesforce, 2022)

In addition to modernizing the skills development strategy, digital curriculum, and skill-based hiring practices, organizations must foster a culture of innovation and lifelong learning. Investments in community-based, social learning programs that promote team-based learning and include the application of various digital ways of working. For example, hackathons, digital learning labs, and community meetups are essential in the digital era and hybrid work environment. In addition, organizations need to consider gamification, which encourages healthy competition and aids employees in tracking their progress. They could also use leaderboards to incentivize new behaviors and support reward and recognition programs. And finally, the proliferation of digital tools and emerging disciplines are multiplying the complexity of the operating environment within organizations, which needs attention.

A NEW BUSINESS AGENDA FOR DIGITAL LEADERSHIP

Businesses grappling with adaptive problems today need to embrace a learning orientation that empowers people to become more adaptable – to experiment, iterate, and innovate and to share the learning around the organization. For many, this will mean finding new ways of working – and of course, overcoming the pushback that inevitably accompanies change.

(London Business School, n.d.)

Effective digital transformations are distinguished mostly by the practices that business leaders choose to follow. These include:

- **Laying out clear priorities**: Focus on a few clear themes tied directly to measurable business outcomes conducive to achieving better results.

- **Investing in talent**: Attract and develop highly talented people with strong digital and analytics capabilities.

- **Committing time and money**: Allocate operating expenditures toward digital transformation priorities that increase the odds of success.

- **Embracing agility**: Revisit and rearrange priorities to meet the breakneck speed at which competitors and customers move in the digital economy. As a result, companies that adhere to agile practices were nearly twice as likely to report that digital-transformation efforts had beat performance expectations.

- **Empowering people**: Foster a shared sense of accountability and create the right operating environment (Nacimiento, 2019).

Digital leaders are focused on advancing capabilities in several key areas, including delivering world-class digital customer experiences, accelerating time to market, reducing inefficiencies while improving product/service quality, investing in, and building a robust digital ecosystem, enhancing employee readiness and experiences, and strengthening talent growth, mobility, and retention.

The digital economy is here to stay, digitalization is on the rise, and it is compelling every single organization on the planet to reimagine what work is done, who does the work, how work is getting done, and where it is getting done. The rise of the digital economy necessitates adopting new digital ways of working and upskilling teams. Upskilling the workforce involves employees learning new competencies to stay in their current role due to the change in skills required or adding certain competencies for career progression. Reskilling, however, requires learning new sets of competencies to transition to a completely new role. Both need to be considered. Yesterday's skills are expiring, new skills are emerging, and needs are evolving. The skills that got us here are not going to get us where the digital economy is headed. Jobs are changing significantly even if people are not changing jobs. Technological progress requires knowledge workers to develop a range of new skills to adjust to the new marketplaces and work environments. The growing skills crisis presents an urgent need for action.

PREPARING A DIGITALLY COMPETENT WORKFORCE

Companies realize that it is increasingly difficult to find talent for digital roles, while demand is high for digital skills, and the supply is low (Feijao et al., 2021). Specifically, it is estimated that 80% of business leaders will use digitalization as a strategic lever by digitizing work processes and offering work-from-anywhere opportunities (The Future of Jobs Report 2020, n.d.). However, organizations are not currently prepared to equip employees to become digitally competent. Specifically, organizations are experiencing the following:

- Proliferation of predominantly uncoordinated digital skills initiatives
- Confusing and conflicting employee development programs that reduce the 'employees' confidence levels
- Unstructured approaches to digital skills readiness
- Unidentified or non-standardized set of digital skills and competences
- Inconsistent learning pathways and curricula
- Fragmented set of digital tools and techniques
- Limited or no hands-on employee learning experiences

The skills development priorities included within a digitalization culture need to include the following core competencies for the digital era:

(a) customer-centric, digital design with a focus on the methodology and framework, such as design thinking, customer journey mapping, and user experiences; (b) business agility, including lean-agile principles and program management practices; (c) citizen-led employee participation through low-code/no-code tools and techniques; (d) real-time, digital collaboration using tools such as Microsoft Teams, Zoom, and Miro; and (e) advanced data analytics, leverage artificial intelligence/machine learning models and tools such as Microsoft PowerBI and Tableau.

Unfortunately, educational institutions worldwide are not currently prepared to equip individuals with digital competencies, resulting in little to no digital fluency. For example, according to the World Economic Forum (2020), outdated educational institutions have contributed to the future worker shortage. Moreover, within the past 10 years, the development of future workers has not grown. Furthermore, organizations have also contributed to the skills shortage. Therefore, closing the skills gap includes developing a robust, modern digital curriculum, education, and training opportunities for all future digital workers.

To close the gap, the authors recommend a framework that is based on a partnership between organizations/business leaders and academic institutions. This framework will include organizations, academic institutions, learners/students, and the community. See Figure 3.2 for a representation of the components of the framework. Specifically, to meet the needs of a digitally competent workforce, including the future worker, there must be a collaboration between humans and digital technologies (including

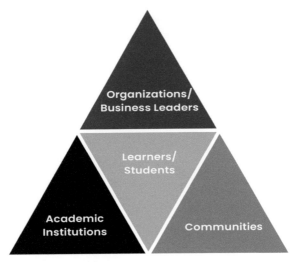

FIGURE 3.2 Components of the Organization and Academic Institution Framework.

machines) for existing job duties (World Manufacturing Report, 2020) and future work. In addition, future work will also require workers to augment their work tasks with future technologies, such as digitalization.

Future workers must adapt their experiences, education, and skill sets to operate and complement the new digital frontier. This must include digital readiness, creating digital awareness, and driving work expectations. See Figure 3.3 for an overview of the digital workforce's partnership

PARTNERSHIP ACTIVITIES

- **Industry Advisory Board**
  Develop advanced digital culture and equitable partnership

- **Advisory Board Cross-Training**
  Train on case studies, user experiences, best practices, and digital content

- **Digital Curricula Development**
  Include curricula content such as design thinking, simulations, user experience user interface, training, agile methodology, and PMBOB methodology

- **Digital Curricula Assessment Review**
  Ensure academic programs are aligned with partnership expectations

- **Ambassador/Mentor Program**
  Work to ensure the success of academic partnerships by collaborating with Faculty Mentors, Campus Champions, and Student Ambassadors

- **Digital Incubators**
  Implement programs where students form teams and solve digital workforce problems such as project management, innovation, design, and bot-a thons

- **Digital Conferences**
  Implement conferences which include resume training, job interview training, interviews with companies, workshops, and panel discussions

- **Academic and Professional Publications**
  Showcase academic and professional partnerships in publications

FIGURE 3.3 Partnership Activities for the Digital Workforce: Organizations and Academic Institutions.

		ORGANIZATIONS	ACADEMIC INSTITUTIONS
		BUSINESS LEADERS WILL:	FACULTY AND ADMINISTRATION WILL:
PARTNERSHIP ACTIVITIES	Industry Advisory Board	Provide advice related to digital readiness, including skills, training, curricula content, tools, and techniques	Prepare students for digitalization, including skills, training, curricula content, tools, and techniques
	Advisory Board Cross-Training	Participate in cross-training with faculty and administration to understand the educational processes and procedures	Participate in cross-training with organizational business leaders to understand the digital community, including the case studies and user experiences that contribute to business value
	Digital Curricula Development	Participate in digital curricula development	Participate in digital curricula development
	Digital Curricula Assessment Review	Participate in the digital curriculum assessment to ensure students graduate with the skills needed to close the digital readiness gap	Develop the digital curriculum assessment processes and procedures
	Ambassador/ Mentor Program	Provide time for ambassador/mentor positions	Select students to participate in the community ambassador/mentorship program
	Digital Incubators	Support digital incubators, including the funding, scope, and digital workforce problems	Support digital incubators, including identifying the students, forming the teams, and suggesting digital workforce problems
	Digital Conferences	Support digital conferences with funding, human resources, job training, and panel presenters	Lead the digital conference execution, including the planning, organization, and implementation
	Academic and Professional Publications	Collaborate to develop academic and professional publications related to partnership	Collaborate to develop academic and professional publications related to partnership

FIGURE 3.4 Specific Partnership Activities for Organizations and Academic Institutions

activities, including organizations and academic institutions. The information included in this figure provides a roadmap that will guide the implementation activities related to the partnership. See also Figure 3.4 to identify the specific activities to be conducted by the organization's

HP INC. SUSTAINABLE IMPACT STRATEGY

HP Inc. has set an ambitious goal to advance digital equity for 150 million individuals by 2030 as part of its Sustainable Impact strategy. To achieve this goal, the company is investing in HP LIFE, a free IT and business skills training program that has already attracted 900,000 learners across 200 countries and territories. HP's digital equity initiatives recognize the critical need to prioritize underrepresented groups, such as women, people with disabilities, communities of color, and educators, as the growing digital divide creates barriers to accessing competitive job opportunities.

HP INC. HBCU TECHNOLOGY CONFERENCE

In 2021, HP Inc. created an inaugural HBCU Technology Conference and first-ever Future of Work Academy (FOWA), focused on faculty, administrators, and students from 100+ US-based Historically Black Colleges and Universities (HBCUs). In partnership with Microsoft, students from minority communities were trained on key trends shaping the future of work. Topics such as Agile Mindset, Design Thinking and Intelligent Automation were introduced, and through a hackathon supported by technical mentors, students took on a hands-on learning experience with low-code/no-code automation bot-a-thon, working in teams, as they looked to solve day-to-day challenges in their student life. Winners of the bot-a-thon interviewed for internship placements at HP, Microsoft and other industry partners participating, as well as met with HP's senior leadership team to showcase their innovation.
(HP Inc., 2021)

MICROSOFT LAUNCHES A GLOBAL SKILLS INITIATIVE

In 2020 Microsoft launched its initiative to help 25 million people worldwide acquire the digital skills needed in a COVID-19 economy. According to Microsoft "one of the key steps needed to foster a safe and successful economic recovery is expanded access to the digital skills needed to fill new jobs ... including people with lower incomes, women, and underrepresented minorities, those hardest hit by job losses"

UIPATH'S ACADEMIC ALLIANCE

In 2019, UiPath's Academic Alliance "A Robot for Every Student" program, has enrolled 1800+ academic instructions, across 68 countries, to offer RPA courses and certification. 2,200 educators around the world have trained 372,000 students enrolled in business and economics, finance and accounting, and other technical programs, opening paths for nearly 50,000 people to new exciting careers in automation.

FIGURE 3.5　Gearing up for a Digitally Ready Workforce through Organization and Academia Collaboration

business leaders and specific activities for the faculty and administrators with the academic institutions.

Organizational initiatives and investments are underway to address the growing digital skills gap and digital inequity challenges. For example, Microsoft launches a global skill initiative, UiPath supports Academic Alliance partnerships, and tech companies are partnering with these global enterprises (Figure 3.5).

CONCLUSION

The shift in the division of labor between humans and machines has led to digital transformation initiatives requiring new digital skills and new digital roles. While the demand for digital skills is high, the supply is low, and businesses are challenged to find individuals with the essential digital skills required. To close the gap, the authors recommend a framework that is based on a partnership between organizations/and academic institutions. The information included in the framework provides a roadmap that will guide the implementation activities related to the partnership.

CHAPTER 4

Intelligent Automation Accessibility and Applicability

Aftab Ahmed

INTRODUCTION

As new technology ushers in the "fourth industrial revolution," perhaps the most striking aspect of this phase in industrial development is the democracy and universality of the tools emerging in the marketplace, allowing everyone to participate. From individuals to businesses, small to big, local to international, across sectors, demographics, and skill sets, these new tools, like IA, allow for remarkable transformations of how humans work.

In this chapter, the authors explore the accessibility and applicability of IA, discussing how inherent flexibility in the pace and nature of learning, mastering, and deploying the tools allows for unprecedented ability to scale at a rapid pace and utilize resources from a wider range of talent pools than ever before. Practical uses of IA are discussed to demonstrate the tremendous adaptability of the tools to address an impressively wide range of scenarios and processes, from the small, singular, and simple to large, multi-activity, and complex. The chapter concludes by reminding the reader that this widespread accessibility and applicability brings in itself additional challenges that have to be overcome, including human fear of change and fear of technology.

DOI: 10.1201/9781003276128-4

IA IS PERHAPS MORE ACCESSIBLE THAN ANY OTHER EMERGING TECHNOLOGY TO DATE

As ubiquitous as the internet, computers, and smart devices have become, the world of software coding has become increasingly integrated into everyday society. For example, today's cell phones contain more technology than the rocket ships that took humanity to the moon 50 years ago, and coding is the key to allowing all the disparate components to work effectively together. But whereas before, the skills to write these codes were limited to those who had studied and applied coding languages for years, in today's world of IA, such barriers to entry are becoming non-existent (see Figure 4.1).

IA offers a low cost of entry, with vendors offering trials of the products for free, free online training, and the ability to achieve certifications to showcase your growing skills. The tool interfaces are deliberately user-friendly, taking advantage of Graphical User Interfaces (GUIs) that the computer age has made us all familiar with and banishing the days of needing to type long lines of syntax. Users are gratified for their efforts with the ability to go rapidly from almost no experience to implementing a solution.

In today's society, where diversity, equity, and inclusion are increasingly the focus of cultural efforts at enterprises across the world, IA empowers all individuals to tool or retool their skill sets to meet the gig economy needs and expectations of the 21st century. IA tools don't care what entity you are working in, as they are process, function, and industry agnostic. As a result, they can take away the mundane tasks from your job, no matter if you are a data processor or a captain of industry.

Learning how to use the tools can begin as early as when children first start using computers or as late as retirees looking to simplify their hobbies. This is no longer the world of university graduates alone having the skills to code.

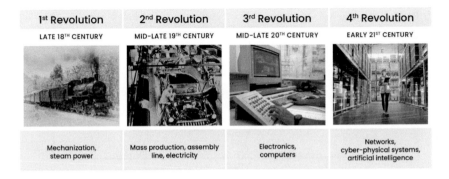

FIGURE 4.1 Industrial Revolutions. (Adapted from Klaus Schwab, World Economic Forum, 1/14/2016.)

And geography, long a barrier to worker mobility, is no longer a limitation. Instead, IA can be done remotely, allowing for opportunities to work from anywhere, opening the doors to new workers and ways of working that are increasingly being embraced in today's post-COVID environment.

IA PROVIDES AN OPPORTUNITY FOR SCALABILITY AT AN UNPRECEDENTED PACE AND SIZE

One of the biggest challenges with introducing any new technology in a business or institutional setting is managing change and scaling the adoption of the new tools. IA tackles this head-on with a model that allows for starting small, perhaps with just one user and one bot, and growing the footprint over time. Flexible license models allow for prioritization and targeting of parts of the organization that will benefit the most and thus trailblaze adoption. Success and resources can be leveraged across business lines, geographies, and time zones because IA is agnostic to all these traditional barriers to scaling adoption.

IA offers a self-serving delivery mode where individuals can build solutions for themselves, their work group, or maybe even their whole company. From almost any corner of the enterprise, such avenues for scaling adoption provide the opportunity for growing use and value at an almost limitless pace and size.

Another advantage of the IA model is dealing with market uncertainties and business cycles. Such uncertainty can cripple businesses that must deal with trying to recruit and retain talent in boom times and lay off and retire workers in downturns. IA can provide for a virtual workforce that can be ramped up and ramped down to meet these business cycles at a fraction of the cost or disruption that traditional worker models necessitate (see Figure 4.2).

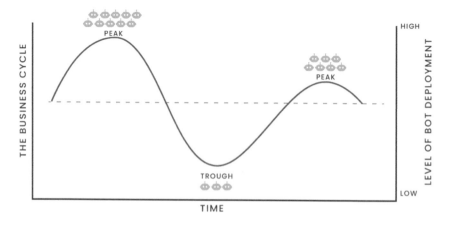

FIGURE 4.2 Using Bots to Scale Up for Business Peaks and Scale Down for Troughs.

IA OFFERS WIDESPREAD CAPABILITIES AND THE ABILITY TO BE USED ANYWHERE

Intelligently automating business processes can seem daunting when you consider the multitude of complexities in modern business, with the intersection of multiple applications, multiple workers, external interfaces, and paper documentation. Historical process improvement efforts have largely focused on reducing inefficiencies within each of these components of a process but rarely were able to revamp the E2E process without requiring major, multi-year project efforts that typically came with major system changes and required a "fitting within the box" acceptance of how efficient the process could really become.

Today's process improvement efforts can take advantage of digital tools to build enhancements as simple or complex as needed based on the inputs, processes, and outputs involved at any stage in the process. In addition, the technology allows for inserting "human-in-the-loop" (actioning when it's their turn) or "human-on-the-loop" (monitoring for intervention when needed) steps into automations to manage risks or maximize efficiencies where needed, and only when needed (see Figure 4.3).

IA often provides the launching pad for wider digital transformation efforts, as well as a center from which complementary tools and cognitive, analytics, and other hyperautomation tools can be coordinated to achieve end business goals. Consider digitally scanned paper documents being read by intelligent Optical Character Recognition (OCR) or Natural Language Processing (NLP) tools to extract the key business data, automatically processed with IA through financial applications and summarized on analytics dashboards for human workers to review for business insights. What was once a multi-day, multi-person, back-dated business

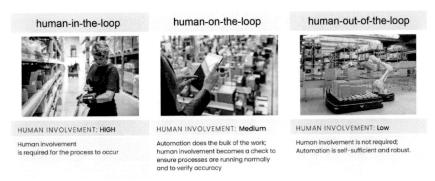

FIGURE 4.3   Human-in-the-Loop vs. Human-on-the-Loop vs. Human-out-of-the-Loop.

activity (when prepping the info took almost all the time) can become a real-time business driver for immediate value insights. Doing more with less, making data-driven decisions, with 24/7 processing and real-time insights, is a North Star for business operations across enterprises as they seek to de-risk their business activities while increasing business value.

The open nature of IA allows the technology to be used with essentially any application, system, or platform that exists today or in the future. This covers back-office functions like accounting and human resources, mid-office groups like supply chain and customer service, and customer-facing front-office workers like traders and sales. There are a myriad of interfaces with different software and hardware, yet all with the potential to be interrogated by, and posted to, by Robotic Process Automation (RPA) IA.

IA CAN BE USED TO IMPROVE ALMOST ANYTHING THAT INVOLVES A WORKFLOW

The adaptability of a tool that can interface with essentially any other tool provides a unique platform from which to automate the things you do on a day-to-day basis. Whether personal, single activity, multiple activities, or even business-to-business, IA offers the ability to take away the mundane and focus on adding value. Let's consider some examples (see Figure 4.4).

Personal Admin

Hands up if you like the bureaucracy of being an employee in a company! My guess is you didn't raise your hand. Nobody likes the form filling, record keeping, and other administrative tasks that companies require of their employees to maintain order, compliance, training and development, and other necessities of business workforce management. For example, consider an office worker recording personal time off (PTO) that most companies require employees to record in other defined timesheet-esque systems depending on the type of time off. Most office workers typically first capture this type of information in their calendars so that they, and others who

| Personal Admin | Daily Desk Activities | Group Workflows | Mutli-Functional/Multi-Business Processes | Business-to-Business Activity |

FIGURE 4.4   RPA Use Buckets.

might try to schedule meetings with a said worker, know when they will be out. Rather than requiring time-consuming double entry in two systems (and the potential for discrepancies, lack of timeliness, or missing entry), IA can take a single entry and post it to other or multiple systems as needed.

Daily Desk Activities

In many office settings today, workers find they have numerous tasks to perform each day, many of which repeat during the day or on a daily, weekly, or another frequency basis. Wouldn't it be great if those tasks could be intelligently automated so you could accomplish the tasks at the click of a button rather than have to repeat the same steps again and again and again? Enter IA, designed specifically to deal with repetitive mundane tasks. For instance, consider a commodity pricing analyst who starts each day by going to multiple websites and opening several emails with reports that show various attributes about the previous day's trading of various commodities. For example, emails about opening, high, low, closing price, the volume traded, etc. Next, the analyst copies or otherwise extracts key data from each site and reports it into a template that is used further to disseminate these highlights to the front office employees. IA can quickly interface with designated websites, scan emails and open reports, intelligently extract the key data points required by the template, and automatically update the template and distribute (with a human-in-the-loop step to approve if needed) once completed. As a result, potential work hours are saved daily, freeing up the analyst to focus on insights vs. data gathering.

Group Workflows

The most striking feature of IA is the manner in which solutions can be connected to allow for more end-to-end process automation. This allows for initially separate solutions to be built, perhaps at a desk-job level, and then sequenced together to crossover multiple roles and steps in a process. For example, consider the process for recording receipt and payment of paper invoices in a small business. One member of the payables group may be responsible for capturing data from the invoice and entering it into the accounting system. Another person on the team may be responsible for matching the invoice to the purchase order and confirming the goods or services receipt. And a third person on the team may be charged with processing and authorizing the invoice for payment. With IA, all three of these workflows can be partly or even fully automated for the three team members and then connected sequentially to allow for end-to-end processing with little to no human involvement.

Multi-Functional/Multi-Business Processes

IA solutions don't have to be limited to one process flow or group's tasks. With the process agnosticism of IA, solutions can be built and connected across functional and business lines to allow different types of work to be automated. A common example of this is in the onboarding of a new employee. Facilities activities usually include badging and officing, IT is typically charged with computer and system setup, and HR often drives the initial compliance and awareness training. Most of these activities typically involve interactions with the new employee's supervisor for approvals and parameters, usually through very disparate systems and applications. IA can deliver solutions to automate the requesting, notifying, and processing of instructions and approvals across these processes and systems, integrating to create a single portal for the supervisor and new employee to see activity status and timing. No more chasing different groups to action steps or missing key steps because of lack of awareness. Every HR department would strive for a much more efficient onboarding process focused on user experiences.

Business-to-Business Activity

Another feature of IA is that the tool can be adapted to fit any type of system or application by focusing on the user interface. This offers the ability to build automations using external interfaces such as websites and third-party applications, bringing data directly into your own systems and applications from the outside. Existing solutions built around electronic data interchange are often costly and limited to just the largest of data flowplace. IA can replicate much of that functionality on much smaller scales for wider coverage. When combined with other emerging technologies like blockchain where the b2b investments have taken place, this opens the possibilities for significant processing efficiencies in b2b activity.

The multidimensional relationships between bots and humans that IA can accommodate, from one-to-one, one-to-many, and many-to-one, offer endless possibilities for how intelligent automation can transform processes, roles, and organization structures to improve efficiencies and capabilities. But, of course, as the complexity and numbers of players increase, so does the need for more robust solution-building and stronger maintenance protocols.

SOUNDS EASY? PERHAPS, BUT THERE ARE STILL PERILS AND FEARS TO MANAGE

So, if IA is truly for everyone, why isn't everyone taking advantage? Well, as with any new technology, it takes time for awareness, opportunity, and results to be seen by enough people to drive widespread adoption.

A primary hurdle for any new technology is "one of fear." As factory automation demonstrated in the early 20th century when assembly lines replaced factory workers, the fear of technology replacing humans and, thus, their livelihoods is a real fear that must be addressed in any company's implementation and at a societal level. History has taught us that the usual refrain of "we will train you in other skills" must be met with real opportunity and practicality. With technology, IA in particular, this is perhaps more possible than ever given some of the themes explored earlier in this chapter, where education, mobility, and age are not the barriers they were for automobile workers, for example, in the last century. But that fear of being unable to learn the new skill or new requirements of their job roles is also very real and must be managed. For example, freeing up workers from the mundane data gathering so they can focus on business insights sounds great in practice, but have they been trained in the skills to identify, analyze, and comment on the insights they are now expected to deliver? Providing more time alone is likely not enough to tool these workers to succeed, and management must plan, recruit and train accordingly.

Indeed, managing the change is perhaps key to delivering the promise that "IA is for everyone." As processes and workflows are changed, the business aspects that go with those changes, from stakeholder experience to business documentation, to roles and responsibilities, to control and to monitor activities, must all be carefully reviewed and updated to reflect the new way of working. For example, supervisors may suddenly be charged with managing "bot workers" – what does this mean, and how do they manage this? In addition, while the technology is perhaps end-user driven and empowering, the role of your IT organization in support and maintenance cannot be ignored and must be clearly defined, established, and tooled to deliver on the business needs and capabilities being relied upon with IA.

There is also the societal aspect to any new technology and how it is being used. Privacy concerns, confidentiality, cybersecurity, and ethical use have a bearing as IA and other digital technologies come to the fore in this fourth industrial revolution. Policymakers worldwide continue to grapple with this as the mega-corporations, small businesses, and consumers make their cases for what they want and need.

So, is IA really going to become for everyone? Alexander Graham Bell said in 1880 of his invention, "One day there will be a telephone in every major city in the USA" (Alexander Graham Bell Quote: "One Day There Will Be a Telephone in Every Major City in the USA.," n.d.). One hundred forty years later, there are an estimated 7+ billion phones in active use around the world, enough for one per person on the planet. Bill Gates and Paul Allen, co-founders of Microsoft, set a goal in the early days of their company in the 1970s for "a computer on every desk and in every home" (Bae, 2015). Fifty years later, we are at 2+ billion computers. Daniel Dines, the co-founder of UiPath, stated his ambition in 2018 of "a robot for every person" (Alexe, 2018). Who would bet against that number being in the billions in a few decades?

CONCLUSION

IA has increased the flexibility and speed at which people can work across a wide range of activities and processes. In addition, these new tools offer the inherent ability to be increasingly adopted at a broader scale and by a wider audience, allowing more and more of society to be included in the significant transformations underway in how people work.

CHAPTER 5

Technology Only Does Not Yield Success

Humble Culture Drives Success in IA

Tiffany Maldonado

Sam Houston State University

ORGANIZATIONS UNDERSTAND THEY NEED to innovate to stay competitive in the ever-changing business environment. However, implementing the latest technological trend isn't enough to drive success. To implement innovation for competitive success, a firm requires supportive actions from leaders, especially when the innovation is complex. To foster complex technological innovations, such as robotics process automation (RPA), firms benefit from having a humble organizational culture. A humble organizational culture has six dimensions: employee development, mistake tolerance, transparency, accurate awareness, recognition, and openness. This culture of humility influences how individuals think and act and how the firm performs and is built and maintained by policies, processes, routines, and practices those strategic leaders nurture.

WHY TECHNOLOGY ALONE ISN'T ENOUGH

Many companies have a false illusion that if they can implement the "right" technology or the latest innovation, their business will succeed. The result

DOI: 10.1201/9781003276128-5

is that companies fail to consider how well the innovation fits with their firm. For example, consider a large bank (Best Bank) with over 100 locations in North America and a medium-sized chocolatier (Sweet Coco) with 25 locations in Europe. These firms will have different innovation needs to help their company grow. For example, both firms may benefit from the latest technology in processing payments; however, Sweet Coco will have little need for RPA that makes repetitive processes, like checking deposits, accessing credit reports, and reconciling accounts. Likewise, Best Bank will have little need for automated inventory management.

Even if we assume that each firm has selected the "right" technology for their firm, the best-case scenario is that these firms will enjoy business growth, but on a temporary basis. All firms seek to have a competitive advantage, which is when consumers pick your firm as opposed to your rivals. The truly successful firm seeks a long-term competitive advantage, also called *sustainable competitive advantage (SCA)*. A firm that achieves SCA is able to enjoy superior financial performance, market share, and brand reputation over its rivals. In order to obtain SCA, firms must be aware that simply implementing technology is not enough. In fact, installing the latest innovation, bot, or AI can only lead to a temporary competitive advantage which will be erased as soon as the firm's rivals implement the same technology. So, what can a company do to achieve sustainable competitive advantage when implementing technological innovations? First, firms must implement and maintain a *humble organizational culture*.

WHAT IS HUMBLE ORGANIZATIONAL CULTURE?

What is a humble organizational culture, and how does it help a firm achieve an SCA instead of a temporary competitive advantage? Humble organizational culture is a type of culture that promotes humility as a source of competitive advantage and consists of six values and behavioral norms: (a) accurate awareness is promoted, (b) competent mistakes are tolerated, (c) transparent and honest communication is rewarded, (d) openness to the ideas of others is valued and modeled, (e) employee development is prioritized, and (f) employee recognition is practiced regularly. Humble organizational culture is nurtured through the firm's policies, processes, and practices. Humble organizational culture has six dimensions (employee development, mistake tolerance, transparency, accurate awareness, recognition, and openness) grouped into three categories: Hindsight, Insight, and Foresight.

Hindsight helps firms to reflect on and understand a situation or event after it has happened. These companies can accurately evaluate past events for overlooked signals to assist in capturing knowledge that can be useful for the future. Insight helps firms to gain a deeply accurate understanding of a person or situation to understand cause-and-effect links and how the present actions of the firm will make an impact. Foresight helps firms to predict what will happen or will be needed in the future to steer the direction and strategy of the firm. Foresight enables firms to combine past and present actions and events to predict which actions will bring success in the future.

HOW DOES HUMBLE ORGANIZATIONAL CULTURE TRANSFORM TECHNOLOGICAL INNOVATIONS?

At this point, you should be familiar with the basics of a humble organizational culture; now, we will discuss how a firm can utilize this culture to transform a technological innovation from a temporary competitive advantage to an SCA (see Figure 5.1).

Through *Hindsight*, companies can use accurate awareness of the strengths and weaknesses of the firm to improve technological innovation to mitigate the weaknesses and capitalize on the strengths. For example, let's look at a medium-sized luxury hotel company (Lux Resorts) with three locations in North America and three in the Caribbean. They will implement a new predictive analytics program powered by artificial intelligence (AI), but they want to ensure that they obtain an SCA by implementing this program. Thankfully the firm has an established humble organizational culture. Lux Hotels uses *Hindsight* to review its performance for the past 3 years. They noticed a decline in the length of stays in the North American locations, and the data suggest that customers in a lower income level stay one day less than the industry average. A team member suggests they can use the firm's customer data to predict which customers would be enticed to stay an extra night, thus spending more money on hotel amenities. The firm decided to tweak the AI program to include a predictive marketing aspect that sends customized *buy five nights-stay seven nights* promotions to select customers. Now that the innovation has been tweaked to mitigate one of the firm's weaknesses, Lux Hotels turns to determine how the same technology can offer upgrades to big-spending customers attracted to the properties.

Through *insight*, firms can use openness and transparency to further improve innovations and stakeholders' ethical treatment. For example,

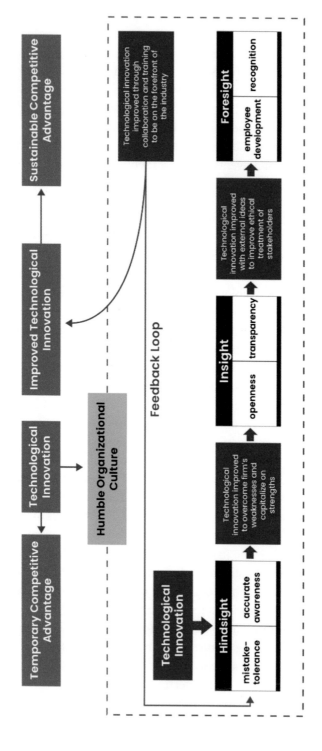

FIGURE 5.1   How Humble Organizational Culture Transforms Technology for Sustainable Competitive Advantage.

Lux Resorts operates through openness and transparency, which results in a team member noting that while the customers at the Caribbean locations enjoy the dedicated customer service line, the service agents do not. Further research reveals that customer agents spend much of their time addressing repetitive issues, such as requests for additional towels, that they are not able to spend the time needed to truly customize experiences for guests. One of the customer service agents suggested that if there was a way to automate or predict which guests will have repetitive requests and how often, agents could use the time saved to work on requests such as romantic dinners by the ocean. Lux Resorts used this external idea to tweak the predictive analytics program to predict when guests will need additional towels based on the number of times and length of showers in the guest room. Based on these data, the housekeepers can drop off additional towels before the guests would need them. As a result, Lux Resorts has improved technological innovation with external ideas to improve the ethical treatment of stakeholders such as employees.

Firms can use employee development and recognition through *foresight* to improve technological innovations to be at the forefront of the industry. Lux Resorts is almost ready to implement the predictive AI analytics program, but they know they should not rush the process. Their employees are key for the next step. After a company-wide survey and feedback from employees who have been recently trained in the latest technology fields, the team revisits the innovation with additional employees in the team. These employees freely collaborate to brainstorm areas of improvement for the program that will cause the firm to stand out from rivals. These employees are confident with sharing ideas due to the humble organizational culture and the value placed on recognizing employees for their contributions. They know their ideas will not be passed off as belonging to management, as Lux Resorts has weekly shout-outs acknowledging employees for their contributions to various projects. Through collaboration, tweaks were made to the program such that every morning during their stay, each guest was greeted with a list of personalized recommendations for amenities based on collected data. This list would provide a list of value-added services guests could purchase and the open slots for such activities based on guest preferences for the morning, afternoon, or evening peak activity levels. Guests would appreciate the truly customized service, and the experience would be akin to having a personal assistant to recommend and book activities. Still, these lists would be generated using the predictive analytics program instead of a human. Lux Resorts is now ready to implement a technological innovation that has been improved through collaboration and training to be at the forefront of the industry. If Lux Resorts decides they are still not happy with the final version or want to update the system, they can follow a feedback loop through the process. By repeating this process with technological innovation, Lux Resorts always stays ahead of rivals and enjoys the benefits of a sustainable competitive advantage.

HOW TO BUILD HUMBLE ORGANIZATIONAL CULTURES

A culture of humility is built when leaders at all levels intentionally manifest their personal humility and reward, encourage, and incentivize humility in the attitudes and behaviors of their team and organizational members. Unfortunately, humble organizational cultures are on a continuum where many companies practice a few of the dimensions of humble organizational cultures. In contrast, few companies implement high levels of humble organizational cultures, which helps these firms to enjoy positive gains regarding firm performance. So, how can you build a culture of humility in your firm, department, or team? The following section will detail how managers can build such a culture to improve outcomes at every level.

Companies wishing to build a humble organizational culture will need to focus on specific actions and changes, not just talk. For example, one company showed *Hindsight* through their actions that they tolerated competent mistakes when an intern sent out an empty test email to thousands of customers. Instead of berating the employee for a mistake that quickly took over social media, the company made it publicly known that they were helping the intern through the mistake. So, what specific actions can a manager implement to start building a humble organizational culture? We have provided a table of very actionable suggestions for managers to build each aspect of humble organizational culture (see Table 5.1).

CONCLUSION

In order to stay at the top of the competition, organizations understand they need to innovate continually. However, as this chapter has shown, simply implementing the latest technological trend isn't enough to drive success. Instead, humble organizational culture drives success, especially when implementing complex innovations. This culture of humility transforms technological innovations into innovations customized to the firm's unique strengths and weaknesses to bring sustained success.

TABLE 5.1 Suggested Managerial Actions to Build a Humble Organizational Culture

Hindsight	Monthly accurate awareness exercise	Challenge a group of direct reports to bring evidence of an external threat (such as a rival gaining market share) and external opportunity (such as the increase of finance graduates) each month. Challenge a different group of direct reports to bring evidence of an internal weakness (such as increased turnover in the department) and internal strength (such as the length of tenure in a department) each month. Compile the lists and circulate them to the team. Then next month, rotate the groups.
	Fruitful mistakes list	To encourage healthy risk-taking, managers should highlight how past mistakes helped to bring forth present solutions. This may take some research, but having a readily available list would encourage employees to make competent mistakes (such as during a toy demonstration, the toy unexpectedly breaks, which leads to the discovery that certain joints do not withstand certain forces)
	Problematic idea box	To increase creativity among employees, install an idea board/box with a sign detailing a particular problem facing the firm (such as the increased turnover in the department). Team members can anonymously submit ideas; at the end of a given period, the (a) riskiest and (b) most creative ideas are highlighted. When implementing this suggestion, only those problems a manager controls and will actually use the ideas to improve the issue should be used. Otherwise, team members will get discouraged, as this is evidence of management talking but not walking the walk.
Insight	Flaw time	During project meetings, reserve some time for the team to find flaws in the current project. Be prepared to suggest the first meaningful flaw if the team is shy. Admitting the weakness in a project takes courage, but it also is the first step toward improvement
	Everyone speaks in focus groups	One way to actively seek new ideas and feedback from multiple team members from all levels is to host focus groups. During these focus groups, feedback is provided about the innovation in process before rollout occurs. The focus group membership should include either one person from every level/division within the group (i.e., administrative assistant, engineer, manager, or North America, South America, and Europe division) or multiple focus groups with the same delineations as above (focus group among the South America division, another group for Europe, etc.)

(Continued)

TABLE 5.1 (*Continued*) Suggested Managerial Actions to Build a Humble Organizational Culture

Foresight	Employee development is prioritized	Focus on employee training
		Encourage difficult goals
	Employee recognition is practiced regularly	Have effective recognition programs in the department. For example, try more innovative awards instead of generic department member of the month. Awards don't have to be formal. For example, after a focus group, you can mention the top three people who contributed the most feedback or select the riskiest idea and acknowledge how it helps to break boundaries
		Be generous with kudos. Encourage an idea board-pose a problem, and employees can post sticky notes of ideas. Then, once a week or bi-weekly, put all the ideas in a raffle and announce a winner. The prize can be formal (such as gift cards, and premium parking) or informal (plastic crown for a day, celebratory die cuts taped to cubicles that can be redeemed for prizes, or not; a gift of a favorite candy) as you like
		Spotlight the achievements of others. Try awarding a plastic lightbulb that rotates to the employee that has contributed the most innovative idea of the month

Source: Adapted from Building a Humble Organizational Culture and Maldonado, T., Vera, D., & Ramos, N. (2018). How Humble Is Your Company Culture? And, Why Does It Matter? *Business Horizons, 61*(5), 745–753. https://doi.org/10.1016/j.bushor.2018.05.005.

CHAPTER 6

Intelligent Automation Implementation in Business

Partha Baral

HP Inc.

Lila Carden

University of Houston

INTRODUCTION

Organizations are leveraging intelligent automation as a tool to meet their strategic goals via digital transformations. These transformations include key factors that drive successful, intelligent automation implementations. We share a deliberate and structured implementation approach, including organizational, operational, management, and measurement models, and the need for diversity in human and non-human resource selections. Additionally, we discuss lessons learned from previous implementations and make recommendations related to phased implementations to support successful transformations and value realization to ensure success.

This chapter will cover how you create readiness, establish an operating model, measure, and deliver value, and operate and govern the digital workforce.

DOI: 10.1201/9781003276128-6

PREPARING FUTURE EMPLOYEES TO USE FUTURE TECHNOLOGY (AUTOMATION)

As noted in Chapter 5 of this book, we use the humble culture approach to prepare the organization, including the employees, for future work related to automation. This human culture approach focuses on the following six dimensions: employee development, mistake-tolerance, transparency, accurate awareness, recognition, and openness (Maldonado, Vera, and Ramos, 2018). The human culture's six dimensions specifically include (a) employee development which is focused on increasing knowledge and encoding it into daily routines; (b) mistake tolerance that supports environments in which decision-making is free from retaliation and embarrassment; (c) transparency related to openness about limitations and acceptance of mistakes; (d) accurate awareness focusing on the assessments of strengths, weakness, threats, and opportunities that are used for strategic decision-making and limitation awareness; (e) recognition to celebrate successes that encourage further challenging opportunities; and (f) openness that encourages new learning that is assimilated into the organization and supports innovation. We also discuss the action plan to prepare future employees to use future technology, emphasizing change awareness and change readiness, including considerations related to **culture, processes, people, procedures, and operations** (Project Management Institute, 2017).

CREATE AWARENESS AND CHANGE READINESS

The *first step* in preparing the organization for a technological change is to make the employees aware of the change. Change awareness focuses on how employees make sense of and the meaning of uncertain change initiatives (Sonenshein & Dholaki, 2012) and needs to consider the dimensions of humility, what to do, what to consider, and who should lead the specific change. (See Table 6.1 for details related to a change awareness implementation plan for automation.)

As organizations move from making employees aware of the change due to automation, the *second step* is to prepare the culture, people, processes, procedures, and operations. Change readiness focuses on preparing the organization and equipping them with the components required to support the change. The more prepared or higher the level of change readiness of the aforementioned dimensions, the more likely the organization will successfully implement the change (Bankar & Gandar, 2012).

TABLE 6.1   Change Awareness Considerations for Automation

Action Plan Criteria/ Dimension	Humility Dimension	What to Do?	What to Consider?	Who Should Lead?
Culture	Accurate awareness	Create the need for change	Strategy, vision, and outcomes	Senior leadership
Processes	Accurate awareness	Create guiding principles	Functional or enterprise implementation	Line managers
People	Recognition	Unify the team, including team-building activities	Recognition for participating as a team member	Human resources
Procedures	Transparency	Communicate the change	Top/down and lateral communications, including forums, blogs, applications, support groups	Communication team
Operations	Openness	Communicate specific impacts on operations	Functional or enterprise operations	Line managers

This change readiness assessment needs to be conducted prior to any change initiative and includes *identification* and *assessment*. For example, the people assessment (as noted in Table 6.1) includes reviewing the skills and capabilities as well as assessing levels of core human needs such as emotional job security, implementation inclusion; being trained to do the new jobs; and perceiving the results of the technological change is fair and equitable (Anderson & Anderson, 2010). People assessment aims to determine the gap between the expectations of change management leaders and the employees implementing and impacted by this change. The organization needs to mitigate the gap so that change resistance will not affect the initiative's success. (See Table 6.2 for the change readiness considerations for automation, including various criteria/dimensions related to humility, what to do, what to consider, and who should lead.)

Establish Operating Model

Everybody is thinking about, reading about, and talking about "intelligent automation" and how to adopt it as a strategic capability to deliver on digital transformation goals. Over the past few years, a tremendous amount of awareness has been created in the industry about RPA (robotic process

TABLE 6.2   Change Readiness Considerations for Automation

Action Plan Criteria/ Dimension	Humility Dimension	What to Do?	What to Consider?	Who Should Lead?
Culture	Mistake-tolerance	Create a humble cultural environment	Honest, creative, and learning environments	Senior leaders
Processes	Accurate awareness	Identify changes to the business processes to support the change	Functional or enterprise implementation	Line managers
People	Employee development	Identify roles and prepare for the future workforce	Training and Retooling	Human resource managers
Procedures	Employee development	Identify the methodology, tools, and techniques that will be used to implement the automation	Agile and program management methodologies	Project managers
Operations	Accurate awareness	Identify technology	Vendors or in-house development	Information technology managers
Operations	Accurate awareness	Determine infrastructure and security needs to support the automation	Vendor and/or in-house changes	Information technology managers

automation), IPA (intelligent process automation), and cognitive process automation. As companies (customers) have started to adopt this, product vendors have invested in maturing the product capabilities/offerings. As a result, best practices for successful implementation and value creation are emerging and being shared.

Irrespective of the company's size, the journey to adopt intelligent automation technology starts small. This is a very typical pattern for success for any emerging technology. This approach of starting small (typically with a proof of concept) ensures success through various required steps of business opportunity identification, technology evaluation, vendor assessment, capturing of unique needs for skills development, resource requirements, technical environment setup, engagement models, governance, communication, and change management, etc.

The journey to adopt intelligent automation and operate at scale goes through three phases: Pilot, Expand, and Scale See Figure 6.1 for a representation of the journey to adoption phases.

The "Pilot" phase involves learning and understanding the market landscape of vendors and solutions and establishing a proof of concept to evaluate vendors and solutions.

The "Expand" phase establishes the foundation of the operating model, architecture, execution, operations, and driving adoption to deliver value.

The "Scale" phase is where intelligent automation becomes an integral part of the transformation strategy, innovation, delivery, and operations of automation solutions performed at scale.

PILOT: Market Assessment and Proof of Concept

Adopting a gradual and thoughtful progression is very important if you consider taking your first step toward exploring intelligent automation. See Figure 6.2 for the steps for evaluation.

The first stage of this progressive journey is to understand and define with clarity what is the goal you are trying to achieve. Answering the questions below will allow you to define the goal:

- Is there a business challenge you are trying to solve? Or is there a business opportunity you want to get ahead of?

- What is the scope/extent of the challenge/opportunity within the enterprise?

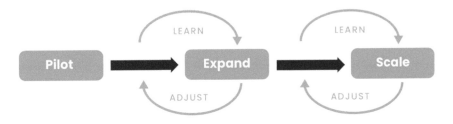

FIGURE 6.1   Journey to Adoption Phases.

FIGURE 6.2   Steps for Evaluation.

- Why is solving this business challenge/opportunity important?

- What impact will the solution have if it can address the business challenge/opportunity?

- Who in the organization is the sponsor for this business challenge/opportunity?

Establish Business Outcome

Define the **business objectives** and the outcome you want to achieve. This is very critical as this will ensure staying true to the end game. All the subsequent steps would need to be guided by this.

Perform Market Research

This is where you need to explore the market canvas and zoom in on what potential solutions and technologies are available.

In addition to the technologies, one needs to explore the following:

- Current maturity of the market (solutions, vendors, investments, etc.)

- How are solutions evolving? (roadmap and strategy for vendor/solution)

- How are companies adopting the technologies? (understand approaches customers are taking to implement these solutions, including success stories, challenges, and lessons learned)

- What use cases are being addressed by the solutions? (which functions, processes, and scenarios customers implement using solutions and technologies)

Key aspects of vendor and technology and products to be evaluated are listed in Table 6.3.

Conduct Proof of Concept

- Define one or more business use case(s) you want to evaluate

- Define the factors to evaluate the potential solutions with

- Define what you want to learn through the exercise

- Establish the right experimentation environment

TABLE 6.3　Vendor and Technology and Product Characteristics

Vendor Characteristics	Technology and Product Characteristics
1. Background and history	1. Current offerings (products and services)
2. Management structure and team	2. Product and services vision and roadmap
3. Holding structure (public or private) and funding structure	3. Product architecture
4. Types and numbers of current customers	4. Security architecture and certifications
5. Customer renewal/retention (how long they have been customers)	5. Integration architecture
6. Partner ecosystem	6. Infrastructure architecture
7. Evaluation by research firms such as Gartner	7. Scalability and configurability
8. Customer references and use cases	8. Licensing and pricing model
	9. Release process and frequency

- Invite solution vendors/implementation partners to participate
- Capture the detailed observations every step of the way using the following dimensions (use as a starting point):
  - Vendor capabilities and processes: how does the vendor guide you through the process – training, learning, product features, engineering, and support processes
  - Architecture and integration
    - Product architecture
    - Integration with other applications
    - Configurability and extensibility
  - Product capabilities and roadmap
    - Technical features (currently available and planned roadmap)
    - Cognitive/AI capabilities
    - Applicability to specific automation use case
  - Development features and experience
    - Usability and experience
    - Engineering skills requirements

- Exception handling

- Configurability of solutions

- Level of effort – implementation

- Deployment and scalability/performance

  - Scalability across the enterprise

  - Usage of telemetry and reporting

  - Operational insight and reporting

  - Level of effort – maintenance and operations

- Security and compliance

  - Authentication and authorization

  - Data security and privacy

  - Integration to privileged access management (PAM) solutions

- Infrastructure

  - Software requirements

  - Hardware requirements

  - Private or public cloud/on-prem

  - Fail-over and disaster recovery

  - High availability requirements and support

- Change management

  - Training and documentation availability

  - Adoption requirements and difficulty

Evaluate Results

- Assess the solutions developed in the proof of concept – compare against the factors/evaluation criteria

- Capture any unique aspects of each solution (even though these may not have been part of the evaluation criteria)

- Compare the results across solutions/vendors/technologies

- If you had multiple use cases, compare the results across these use cases (what unique characteristics did they highlight)
- Validate the results against the intended business outcome defined earlier

Document Learning

- Conduct a retrospect to understand what worked, what did not, and what is needed to make it a success
- Review the detailed learning – readiness, engagement, technology, skills/experience, intended outcome
- Determine what the next stage of the journey should be

This becomes the precursor to establishing the foundation to build an ongoing operating model.

Expand: Expansion Strategy

However, depending on your organizational operating model and culture in approaching digital transformation, you may find yourself in one of the following situations noted in Figure 6.3, and Table 6.4 presents the example scenarios.

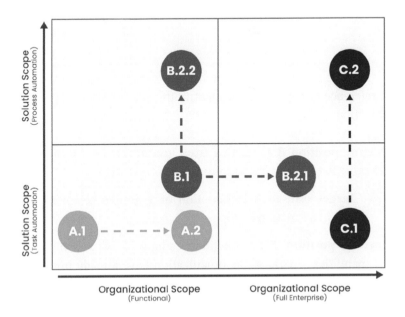

FIGURE 6.3   Expansion and Scaling.

TABLE 6.4   Examples of Scenarios – Organizational and Solution Scope

	Solution Scope (Task Automation)	Solution Scope (Process Automation)
Organizational Scope (Functional)	RPA solutions automate discrete tasks or sub-process activities within a specific functional scope. **Example:** Automation of discrete reporting or data manipulation/reconciliation tasks within finance leveraging RPA solution capabilities.	Integrated solutions (including multiple technologies such as RPA, OCR, AI, and NLP) to automate end-to-end processes within a specific functional scope. **Example:** Automation of end-to-end cash applications or invoice processing within finance requires integrated solutions that include RPA, NLP, OCR, and process mining
Organization Scope (Enterprise)	RPA solutions automate discrete tasks or sub-processes across many functions within the enterprise. **Example:** Many functions (e.g., finance, sales operations, customer support, legal, and procurement) adopt RPA to automate discrete tasks or sub-process activities.	Integrated solutions (including multiple technologies such as RPA, OCR, AI, and NLP) to automate end-to-end processes within and across functions. **Example:** Automation of procure-to-pay (procurement and finance) or order-to-cash (order management and finance) processes requiring integrated solutions that include RPA, NLP, OCR, and process mining

In reference to Figure 6.3, we identify the various starting positions of organizations.

Starting position A.1: Starting from ground zero – An organization begins the journey from ground zero.

- **Target shift to A.2**: Adoption of a limited set of capabilities within a functional scope

Starting position B.1: Functional adoption of a limited set of capabilities – parts of the organization have already initiated their journey without the structured approach.

- **Target shift to B.2.1**: Adoption of a limited set of capabilities within additional functions or

- **Target shift to B.2.2**: Adoption of an additional set of capabilities within the same functional scope

Starting position C.1: Enterprise adoption of a limited set of capabilities – parts of the organization have fully adopted intelligent automation but within a narrow functional scope

- **Target shift to C.2**: Adoption of an additional set of capabilities across the enterprise scope

Depending on your starting position (A.1 or B.1 or C.1) and your organizational strategy and operating model, you may follow a different trajectory to expand automation adoption.

In addition to determining the right strategy to expand the automation footprint, it is also important to establish an "operating model to drive the acceleration of automation adoption" within the organization. This automation operating model will enable the right structure, governance, operating cadence, value delivery, and measurement.

The various aspects of the operating model start with establishing a clear vision, strategy, and outcome for adopting IPA for the organization (see Figure 6.4). In order to realize the vision through the execution of the specific strategy and achieve the desired outcome, the following six components are to be defined, established, and matured:

1. **Organizational capabilities:** Identity, define, and establish the right organizational capabilities needed to get started and a plan to develop and mature over time.

2. **Operating model and governance**: Establish an operating model and governance structure that enables the right level of agility, speed, rigor, oversight, and alignment.

3. **Technology, infrastructure, and security:** Define a technology ecosystem architecture, a process to evaluate and incorporate emerging technologies in the ecosystem architecture to deliver on the automation strategies and keep pace with industry innovation. An explicit infrastructure strategy (cloud vs. on-prem vs. hybrid) and a "design for security" principle must be incorporated per the organization's IT infrastructure and cybersecurity requirements and policies.

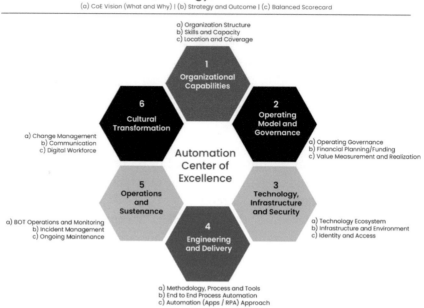

FIGURE 6.4   Automation Operating Model.

4. **Engineering and delivery**: A well-designed engineering and delivery approach is critical to providing automation solutions at scale. The approach must incorporate various engineering processes to enable standardization, repeatability, quality, and scale.

5. **Operations and sustenance**: A well-structured and enabled strategy to operate and maintain the "digital workforce" is a critical component to scale the adoption and stability of intelligent automation capabilities. The right focus, skills, and model for operations and sustenance of the "digital workers" ensure the realization of intended business value and the end-to-end life cycle of intelligent automation solutions.

6. **Cultural transformation**: Digital transformation, when done right, focuses on transforming people (employees are the most precious resource in a company) and the culture to adopt a new way of operating and delivering customer value. It is in the same vein that the intelligent automation program needs to focus on how to bring the employees of the company along the journey to learn, change, and adopt intelligent automation capabilities.

A comprehensive approach and consideration of these six components and vision, strategy, and outcome within a specific organizational context is critical to scale the adoption of IPA capabilities. The details are outlined in the next section, "Scaling Strategy."

Scale: Scaling Strategy

The strategy to drive accelerated adoption of automation at scale must incorporate a comprehensive approach to enabling various aspects, starting with establishing a clear vision and strategy to drive the cultural change. See Figure 6.4 for the automation operating model detail.

It is of utmost importance to define and articulate a clear vision of why an organization wants to invest in and adopt intelligent automation capabilities. While it is very easy to get started with automation and exciting to implement a few initial automation solutions that demonstrate the promise – taking the time to create, socialize, and confirm the vision with senior leaders and various stakeholders within the organization is critical for success.

A vision statement must address the "what" and the "why" of investing in intelligent automation capabilities. The vision statement must also be simple enough for the broader stakeholder groups and employees to relate to and understand. In addition, a clear vision must help drive clarity in decision-making for the program.

It is absolutely critical to establish the primary drivers for implementing and investing in intelligent automation capabilities (Seasongood, 2016; Lacity & Willcocks, 2016; Maldonado et al., 2018; Carden et al., 2019):

- **Cost savings and avoidance**: Automation reduces manual work and hence requires fewer human resources. This is one of the easiest and most common drivers for investing in automation. However, this is also one of the drivers where change management becomes very critical to appropriately address the apprehension from the employees as this is viewed as taking jobs away.

- **Innovation and agility**: Automation capabilities are critical enablers for innovation and agility. While enterprise IT applications (systems of record, systems of engagement) are the platforms to run the core business processes to deliver on the foundational data and experience requirements, intelligent automation capabilities can be leveraged to implement innovation. The innovation includes automated processes as well as speedy, personalized customer/partner/employee experiences.

Intelligent automation capabilities are critical for innovation and agility. This enables an organization to design and deploy innovative processes and experiences iteratively and faster. The intelligent automation capabilities interface with enterprise IT applications and platforms to deliver speed, agility, and quality solutions.

- **Revenue and margin improvement**: Automation capabilities are leveraged to focus on enabling processes that drive (directly or indirectly) revenue and margin improvements. Typically, these processes are related to sales, product introduction, new business models, and/or finance processes dealing with revenue and margin analysis, generation, and recommendation.

- **Customer and partner experience**: Automation capabilities are focused on the processes that involve customer and partner interactions. Intelligent automation capabilities are leveraged to simplify and automate processes and information sharing to create a better and more personalized experience for customers and partners.

  This can also be achieved by creating automation for employees to deliver memorable and effective interactions with customers and partners. Examples of areas of focus are customer or partner-facing roles such as sales, customer care/support, order management, service delivery, field support, customer/partner operations, account receivables, claims processing, and billing.

- **Employee productivity and experience**: Focusing automation capabilities on improving employee productivity and experience provides companies with a huge advantage in maximizing the value of human skills. Freeing human resources from non-value-added repetitive tasks and reskilling human resources to perform higher-value activities allows companies to invest their human capital fully. In addition, this also enables a company to prepare the workforce for the "future of work" through effective reskilling.

The process of sharing and aligning on the vision needs to be iterative and inclusive for understanding/incorporating the feedback from the stakeholders. This will ensure participation and buy-in through the vision's co-creation and the stakeholders.

There need to be clear strategies to realize the vision and intended outcomes based on the vision. The strategies to realize the vision include the company's culture, the digital transformation approach, and the organizational operating principles.

Based on the drivers for value creation applicable to a company, the intended outcomes must be defined to align with the drivers. This will help both in the measurement of progress and the value delivered. The measurement metrics or KPIs and other operational performance metrics can be prepared, tracked, and communicated in the form of a Balance Scorecard across the stakeholder groups. This facilitates transparency, collaboration, partnership, and inclusiveness across the organization to mature and scale automation capabilities.

Digital Workforce – Integrating into the Organizational Model

One critical aspect of formulating the strategy is to first decide on the approach for "automation" – how we define the digital workforce strategy and how it is the same as or different from "traditional IT automation."

We have been observing a shift (imagining the future of work) in the "definition of work," including:

- **What** work gets done to create value (the definition of work and value)

- **Who** does the work (inverting the labor pyramid – human worker and/or digital worker)

- **Where** work gets done from (geography/location is becoming less critical)

- **How** work gets done (adopting an algorithmic model of learning and decision-making)

"Traditional IT automation solutions" are broadly defined as creating systems of record and systems of engagement. And these systems are traditionally implemented and managed by the IT department in a company. However, the advent of RPA, chatbots, AI/ML algorithms, and no-code/low-code technologies enabled every company to adopt a new automation model (in addition to the IT-driven automation) leveraging these capabilities. The "new automation model" allows companies to enable department, team, and individual-level automations (either discrete or integrated) to be created and managed through a centralized, decentralized, or federated governance structure.

The notion of "digital workforce" is defined as the creation of robots (using RPA, chatbots, AI/ML algorithms, no code/low code technologies) known as digital workers and how to integrate these digital workers into the company's workforce strategy (Colbert et al., 2016).

Digital Worker – Digital Twin of the Employee

To create and realize the potential of the "digital twin of the employee" (a.k.a. digital worker), we have to first characterize the human worker definition in a way that can then be translated to create the "digital twin." A human worker is assigned to a specific **role** within an organization and has a specific set of **responsibilities** (what activities are performed and what value is created). The human worker also brings a specific set of **skills** (and develops additional skills) to execute the activities/deliver value in their role. We can codify these roles, responsibilities, and skills to create the "digital twin" of the employees and employ these digital workers to get the job done. See Figure 6.5 for a representation of augmenting human work and the workforce.

We can codify the "skills" of a human worker in a few categories such as (a) task execution capabilities – ability to execute repetitive activities (just get it done); (b) sensory capabilities – ability to see/interpret unstructured

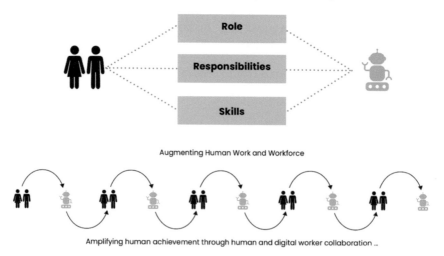

Augmenting Human Work and Workforce

Amplifying human achievement through human and digital worker collaboration ...

FIGURE 6.5   Augmenting Human Work and the Workforce.

information, and ability to feel intent/emotion/sentiment; and (c) decision-making capabilities – learn and make a judgment based on past experience/situational awareness/available information. See Figure 6.6 for comparable skills codification between a human and a digital worker. Also, Table 6.5 highlights typical automation examples. Moreover, we can now combine

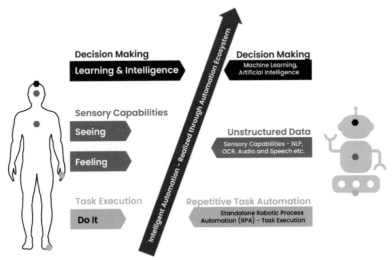

FIGURE 6.6 Digital Workforce – Augmenting Human Capabilities.

TABLE 6.5 Example of Automations and Other Information

Types of Automations	Examples	Key Characteristics
Automations that **Do**	Integrating data between systems generates reports, enters data into transactional systems, auto-populates web forms, etc.	Basic RPA – Automate swivel-chair tasks, repetitive tasks, rule-based activities dealing with structured data and predefined rules
Automations that **Understand**	Invoice processing, sentiment analysis of emails, case notes, data extraction from contracts, etc.	Integrate RPA with OCR, NLP, and audio/speech recognition ability to process unstructured data, and understand intent, language, image, voice, etc.
Automations that **Learn**	Context-aware conversational/voice-enabled BOTs powered by NLP, deep learning deployed in inbound/outbound contact centers, etc.	Integrate RPA with machine learning, deep learning capabilities

and integrate various advanced technologies (RPA, NLP, OCR, computer vision, audio and speech recognition, AI/ML models and algorithms, etc.) to design the relevant skills and shape the digital persona of the human worker. And this enables us to employ these digital workers with the necessary skills to perform the responsibilities/activities in each role.

Workforce Strategy – Optimal Mix of Human and Digital Workers to Optimize Value Creation

To realize the vision of a **fully digitally enabled operating model**, we have to digitize the business processes, including customer interactions/ engagements (digital twin for the enterprise), and integrate the "digital worker" (digital twin for the employee) into our workforce strategy. By incorporating and integrating the "digital worker" into our organizational structure and workforce strategy, we can fully optimize what work needs to get done and how to deliver the work/outcome through an optimal mix of human and digital resources (Phillips & Collins, 2019). For example, see Figure 6.7 noting the integrated workforce.

While we can capitalize on the efficiency advantages and cost advantages of delivering the work by integrating the digital workforce with our human workforce, the bigger benefits of this approach go far beyond cost optimization:

Integrated Workforce – Human and Digital

How Work Gets Done

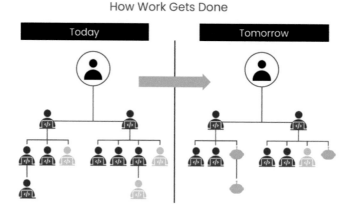

Full Time Employee
Hired by the company as "full time employee" and part of the company's headcount.

Alternate Worker
Hired by the company for temporary or engagement based work and not included in company's headcount.

Digital Worker
Software Robots built using RPA, AI/ML, Chatbot technologies that can perform human like activities.

FIGURE 6.7 Integrated Workforce – Human and Digital.

- Improve customer/partner experience through delivering always-on, individually personalized, and easy-to-do business with engagements and interactions

- Improve the employee experience by shifting the leverage to value-added work for the employees

- Improve accuracy, speed, and efficiency in the processes

- Improve agility and speed to react to changing market dynamics

- Foster business innovation with a focus on time to value

Organizational Capabilities

Organizational capabilities are the components of the foundational structural framework needed to scale the adoption of intelligent automation across the enterprise. It enables and ensures: (a) alignment of value creation through intelligent automation to the overall transformation strategy and (b) effectiveness of the automation program to deliver the intended return on investment. This involves establishing a clear definition as well as an alignment across the enterprise on (a) organization structure, (b) skills and capacity, and (c) location and coverage.

Understanding your dynamic organizational capabilities is an essential component of the foundational structural framework required to scale the adoption of Intelligent Automation across the enterprise.

Organization Structure

Defining and establishing a center of excellence (CoE) model and associated organizational structure is critical within any organization to scale adoption of intelligent automation because of the following:

a. The intelligent automation landscape continues to evolve regarding the vendors, offerings, solutions, etc.

b. Emerging technologies categorized within the intelligent automation ecosystem are maturing

c. Companies are figuring out how to adopt these innovative capabilities and scaling the value creation

A CoE is needed to establish governance and support the rapid deployment of intelligent automation solutions through capability building, training and certifications, architecture and standards, vendor management, solution delivery, operations, and the creation of a library of reusable solution patterns (Anagnoste, 2018; Noppen et al., 2020; Marciniak & Stanisławski, 2021). However, it is also important to establish an "overarching operating model" (see next section) where the CoE is one of the components.

An intelligent automation CoE is a centralized team with a specific focus on driving the adoption of intelligent automation capabilities across an organization. The team size, structure, and skills of resources depend on many factors, such as organizational structure, speed, scale, and maturity of the environment. The CoE structure enables an organization to build a leverage model by delivering a set of core services for:

- Establishing, maintaining, and operating the Intelligent Automation Ecosystem Platform that enables the entire organization to leverage and adopt

- Defining and driving adoption of technology ecosystem architecture and standards (including infrastructure architecture, security standards, data standards, and emerging technology evaluation/adoption)

- Development and consumption of knowledge/expertise

- Sharing and adopting best practices and standards

- Vendor and investment management (cost control by reducing duplication)

- Designing and delivering intelligent automation solutions

- Operating and sustaining the "digital workers."

- Developing and enabling a common set of processes and practices for company-wide programs such as citizen-led development, automation community, etc.

There is a great deal of debate and discussion about where the intelligent automation CoE fits within the organizational structure. However, the success of the organizational setup depends on a few factors:

a. Organizational structure and accountability (centralized or decentralized, or mixed)

b. Strategic objective setting, investments, and prioritization decision-making processes digital transformation strategy (how and which function drives the strategic focus for digital transformation)

c. Role of IT (what role does the IT department play in the company and digital transformation)

The three most common models for the organizational alignment for the intelligent automation CoE are:

1. Chief Digital Transformation Officer

   - Drives the strategic focus for digital transformation; this is the most appropriate organization for the intelligent automation CoE.

   - Ensures close alignment of intelligent automation focus to the overall digital transformation strategy and agenda.

   - Enables the right prioritization of automation opportunities by leveraging intelligent automation capabilities.

   - Establishes the right level of governance and accountability for value realization through intelligent automation.

2. Chief Financial Officer

   - Manage investment decisions, planning, and governance through the Finance organization (the intelligent automation CoE can be part of the Finance organization enabling company-wide automation strategy and focus.)

   - Adoption of RPA typically begins in the Finance organization. Historically, finance processes were naturally transformed to take advantage of automation capabilities to reduce costs and modernize finance operations. It then becomes a natural expansion of intelligent automation CoE's scope to deliver automation capabilities to other functions across the entire company.

   - However, one potential downside can be that "cost savings" (and headcount reduction) become the primary driver for automation. This creates a very narrow focus and minimizes the potential value creation through automation. In addition, "cost savings" (and headcount reduction) as a primary driver of automation can also introduce a "fear of job loss" within the employees and hence impede success.

3. Chief Information Officer

- The IT department is also a natural fit for the intelligent automation CoE. As a part of the IT scope and focus within a company, the intelligent automation capabilities, along with a broader focus on emerging technologies, can naturally become enablers of business innovation.

- **Technology investment and planning**: Given that the IT department owns and manages the technology portfolio along with investment, planning, and implementation, this setup allows for the intelligent automation capabilities to be implicitly part of and closely aligned with the overall technology platform strategy and investment.

- **Architecture and security**: The enterprise architecture focus within an IT department defines, enables, and standardizes an architecture blueprint to implement the company's systems of record, engagement, and innovation systems. The integration of the intelligent automation CoE within the IT department allows for intelligent automation capabilities to be part of this overall architecture standard and blueprint. In addition, this facilitates a deliberate integration of cybersecurity policies and requirements into intelligent automation capabilities.

- **Infrastructure and operations**: Aligning within the IT department also benefit from integrating into its overall infrastructure strategy and operations framework.

See Figure 6.8 noting a typical intelligent automation CoE, and Table 6.6 noting focus areas and their associated detail.

Once the organizational structure for the intelligent automation CoE (Figure 6.8) is aligned, we need to establish the exact skills, resources, and capacity we would need to build the CoE (Table 6.6).

Skills and Capacity

The scope, maturity, and scale of the intelligent automation CoE within the company drive the exact skills and capacity of resources within the

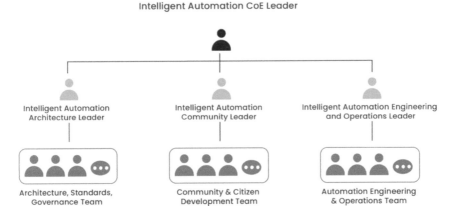

Intelligent Automation CoE Leader

Intelligent Automation Architecture Leader

Intelligent Automation Community Leader

Intelligent Automation Engineering and Operations Leader

Architecture, Standards, Governance Team

Community & Citizen Development Team

Automation Engineering & Operations Team

FIGURE 6.8   Typical Intelligent Automation CoE.

CoE. Table 6.7 defines a broad set of skills required in a fully mature intelligent automation CoE operating at scale. The column "Critical to Start" identifies the skills required to get started.

One of the key considerations for establishing a capacity model is defining an approach for integrating an implementation and operations partner. Depending on the need, a partner typically can provide support for engineering skills, flexible capacity, and operational support.

Location and Coverage

Location and coverage strategies for resources within the intelligent automation CoE become critical factors in scaling automation capabilities across the enterprise. The approach to determine the location and coverage plan needs to be aligned to (a) organization size, geographical spread, and operating model, (b) scope and scale of automation capabilities, and (c) operational and support requirements.

Outlined below are a few key factors to consider for location and coverage strategy:

- **Geographical spread**: Global or focused on a certain geography.

- **Size and scope**: The size of the team depends on the scope and speed – a smaller team may need to be in one location, vs. a larger team can be distributed across multiple locations.

- **Operating model**: Concentration of operational teams in a certain location may warrant co-locating the intelligent automation team to gain a proximity advantage.

TABLE 6.6   Focus Area and Associated Details

Focus Area	Details
Automation Architecture, Standards, and Governance	*Accelerate intelligent automation adoption at scale through the right architecture, standards, governance, and infrastructure.* This team typically consists of architects who focus on <ul><li>Understanding the emerging technology landscape, industry, and technology evolution</li><li>Establishing intelligent automation ecosystem architecture and standards for technology platforms and infrastructure</li><li>Evaluating and selecting any new emerging technologies to be part of the ecosystem architecture to deliver integrated solutions</li><li>Driving adoption of architecture and standards, governance of investment planning, automation adoption, license management, value measurement, and realization</li></ul>
Automation Community and Citizen Development	*Drive automation democratization within the company by building, enabling, leading, engaging, and energizing the community of citizen builders, contributors, and consumers.* This team focuses primarily on community development, engagement, and driving the citizen development program and value creation through the democratization of automation. <ul><li>Building, enabling, engaging, championing, and energizing the community of automation enthusiasts, practitioners, developers, and consumers</li><li>Running community events such as hackathons and learning labs to educate, train and mobilize citizen development</li><li>Building training content, how-to-get-started guides, learning paths, curriculum, etc.</li><li>Evangelizing automation democratization concepts, coaching/mentoring citizen developers, promoting/sharing citizen development success stories</li></ul>
Automation Engineering and Operations	*Drive agile factory-style execution model for delivering and operating cognitive automation solutions.* This team focuses on developing, delivering, and operating the digital workforce to automate business processes and manual activities. <ul><li>Automation ideation, assessment, prioritization, and planning</li><li>Developing and delivering automation solutions</li><li>Developing reusable automation solutions, components</li><li>Operating the digital workforce – managing, monitoring the automations in the production environment, providing operational and issue resolution support</li></ul>

TABLE 6.7   Role/Skills CoE

Role/Skills	Focus	Critical to Start
Intelligent Automation Leadership		
Intelligent Automation CoE Leader	Manages the Intelligent Automation CoE, defines the automation strategy, and acts as the company's Intelligent Automation evangelist	Yes
Intelligent Automation Architecture, Standards and Governance		
Automation Architecture Leader	Leads the architecture, standards, and governance focus of the Intelligent Automation CoE. Establishes the direction for the team, a vision and roadmap for the to-be architecture for the automation ecosystem, and drives governance and adoption of the architectural standards across the company	Yes
Automation Ecosystem Architect	Defines the architecture of the intelligent automation ecosystem that includes RPA, adjacent technologies such as OCR, AI/ML, NLP, and chatbot, and integration approaches to include other emerging technologies. Defines the architecture and integration standards and drives adoption of the same within the company (e.g., Intelligent Automation CoE, Functional CoEs, and Citizen Developers). Partners with technology vendors to understand product roadmap and incorporate it within the automation ecosystem architecture	Yes
Automation Infrastructure Architect	Defines the infrastructure architecture design for the intelligent automation ecosystem to meet the need for different environments, data, and integration requirements. In addition, the infrastructure architecture design needs to inherently account for scalability, performance, robustness (availability, redundancy, etc.), and security of the automation capabilities	Yes
AI/ML Architect	Integrating artificial intelligence and machine learning models in the automation ecosystem is critical for fully automated end-to-end processes and workflows. Defines the architecture of integrating AI/ML capabilities within the Intelligent Automation ecosystem architecture and how automation can be combined with AI/ML capabilities to deliver end-to-end solutions. Partners with automation and AI/ML vendors to understand product roadmap and incorporate it within the automation ecosystem architecture. As the AI/ML domains can be broad (e.g., natural language processing, conversational AI or Chatbot, classification, forecasting, computer vision, and document/image processing), specific skills may be needed in some of these areas.	No

(Continued)

TABLE 6.7 (*Continued*)   Role/Skills CoE

Role/Skills	Focus	Critical to Start
Intelligent Automation Community and Citizen Development		
Automation Community Leader	Leads and drives the focus to build, engage and mature the automation community within the company. Champions the democratization of automation through citizen-led automation programs. Responsible for building learning programs and content as well as community events	Yes
Automation Community Manager	Defines, executes, and measures the success of various community programs across the company. These community programs can be community meetings (e.g., community get-togethers), learning events (e.g., learning labs), ideation/ solution development events (e.g., hackathons), etc. In addition, it defines and monitors automation adoption metrics and community engagement metrics and adjusts programs based on the learnings	Yes
Automation Evangelist	Evangelizes automation technologies, promotes automation adoption, champions and supports citizen developers, drives awareness and sharing programs, and partners with technology vendors to understand product roadmap and communicate back within the company	No
Automation Change Manager	Change management is a critical component of the successful adoption of the democratization of automation and the "future of work" strategy. Defines and executes the change, communication, and engagement strategy and approaches for different stakeholder groups (based on organization levels, roles, functional domains, geography, etc.) within the company.	Yes
Automation Engineering and Operations		
Automation Engineering and Operations Leader	Leads and manages the automation engineering and operations focus on delivering the automation capability roadmap (building reusable automation solutions, components, etc.), executing automation projects, creating solutions, and operating and sustaining the digital workforce. It also drives focus on creating and enabling a set of engineering practices and automation platform capabilities to support both CoE-led automation engineering and citizen-led development	Yes
Automation Engineer	Designs, develops, and tests the automation solutions (including integrating automation solutions to the enterprise applications in the most scalable and effective way) to leverage the automation ecosystem platform capabilities (e.g., RPA and other emerging technologies such as OCR, NLP, AI/ML, and chatbot)	Yes

(*Continued*)

TABLE 6.7 (*Continued*)   Role/Skills CoE

Role/Skills	Focus	Critical to Start
AI/ML Engineer	Designs, develops, and tests the AI/ML solutions/models and integrates the automation solutions leveraging the automation ecosystem platform capabilities. As the AI/ML domains can be broad (e.g., natural language processing, conversational AI or chatbot, classification, forecasting, computer vision, and document/image processing), specific skills may be needed in some of these areas.	No
Project Manager/ Scrum Master	Manages intelligent automation solutions implementations and delivery for the CoE	Yes
Automation Business Analyst	Functions as a process designer, conducts process analysis, documents, and re-engineers processes for automation with a deliberate focus on driving tangible business outcomes such as customer/partner/employee experience improvement, revenue/margin increase, cost avoidance/savings, and compliance/quality improvement	Yes
Automation Operations Supervisor	Supervises and manages the digital workforce's administration, operation, and controls, executing the business process steps/ activities in an automated way. Also responsible for deploying new solutions (digital workers) into the production environment, managing/monitoring utilization, communicating operational highlights to the stakeholders, enabling business continuity planning, and ongoing sustenance of the digital workforce	Yes
Automation Service Analyst	Serves as first-line support for automation solutions	Yes

- **Operational and support requirements**: Global presence may require 24×7 support and operational monitoring, hence the need for a geographically distributed team to ensure coverage. In addition, business continuity requirements may drive the team's presence in multiple locations.

- **Cost optimization**: Cost drivers may need to be considered to account for a location strategy for cost optimization.

- **Implementation and operations partner**: The location and presence of implementation and operations partner resources may need to be considered to create an optimal mix of internal employees and partner resources to achieve the desired outcome.

- **Special requirements**: Certain legal and regulatory requirements may require specific considerations for the team's location (e.g., automation processes for US Federal customers may require the team me

Operating Model and Governance

An operating model and governance structure are critical to enabling the right level of agility, speed, rigor, oversight, and alignment. The overarching automation operating model defines how the automation vision for the company is realized, enables alignment and governance of the execution, measures progress and value delivery, and oversees the financial planning and investment model.

The primary elements of the operating model and governance are:

- Operating governance

- Automation governance structure

- Roles and responsibilities

- Prioritization and decision-making

- Financial planning and funding

- Value measurement and realization

Operating Governance: Automation Governance Structure

The purpose of an operating governance model is to oversee the execution of the automation strategies, ensure progress, realize the automation value, and enable financial and organizational support.

To achieve this purpose, you must first establish an "automation operating committee" representing cross-functional senior leadership across the company. This committee must include representation from various business functions within the company, such as Finance, Legal, Human Resources, Sales, Marketing, Operations, Supply Chain, and Customer Support. The opportunities for automation come from either functional or cross-functional processes within these functions.

It is also critical to ensure representation from supporting and enabling teams to drive the adoption of automation capabilities and advise incorporating specific security, audit, workforce, architecture, and technology requirements. These supporting teams include:

- **Information technology**: It ensures alignment of automation capabilities to the overall technology strategy.

- **Enterprise architecture**: It ensures incorporating automation platform into the enterprise architecture.

- **Cybersecurity**: It ensures the designing of automation platform infrastructure, architecture, and automation solutions with security requirements into the overall automation framework.

- **Internal audit**: It ensures incorporating the right types of controls (business, process, segregation of duties, etc.) are implemented into the automation framework.

- **Human resources**: It ensures alignment to human resources policies/regulatory requirements and integrates the "digital workers" into the overall workforce strategy.

In addition to the "Automation Operating Committee," there may also be a need to establish an "Automation Champions Group (Line of Business)" to drive execution-level planning, decision-making, and alignment. This automation champions group must include representation from each line of business/function, including internal Audit, enterprise architecture, data governance, cyber security, program management office, etc. See Figure 6.9 for the decision-making flow across the automation-focused roles and functional representation.

Refer to the "Operating Governance: Role and Responsibility" section for a detailed explanation of the key focus for each of these groups.

The operating structure and framework must incorporate the automation life cycle's distributed (some may call it "decentralized") nature. There are two aspects of the "distributed nature of automation life cycle" that needs to be considered in the operating framework:

- Automation democratization/citizen-led development

- Federated/distributed delivery model

Automation Democratization/Citizen-Led Development
As low code/no-code automation platforms gain maturity, we must prepare the workforce for what lies ahead, the "future of work." The democratization of automation through "citizen-led automation development" is a

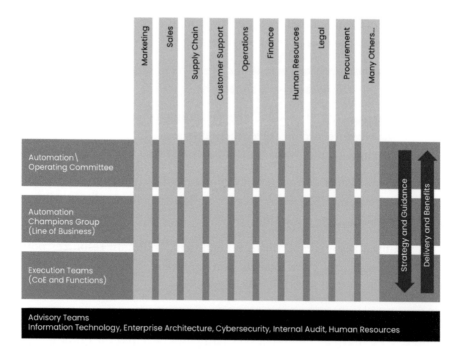

FIGURE 6.9   Operating Structure.

key strategy to accelerate digital transformation through scaling automation adoption and value creation. This preparation enables the organization to accelerate automation scaling and reskilling the workforce for the future See Figure 6.10, including the automation opportunities graph.

The citizen-led development model introduces additional elements to consider in designing the operating framework:

- **Automation opportunity and management**: Need to consider the process for ideation, automation opportunity capture, prioritization, and management work to enable the citizen developers to automate processes (either on a personal level or at a team level).

- **Licensing and adoption of standardization**: Need to account for specific licensing needs (and hence budgeting for licensing cost) and monitor the citizen developers' adoption/usage of these licenses.

- **Value capture and monitoring**: One of the critical aspects of the operating framework is to establish a value capture and monitoring model. This is especially important to consider citizen development

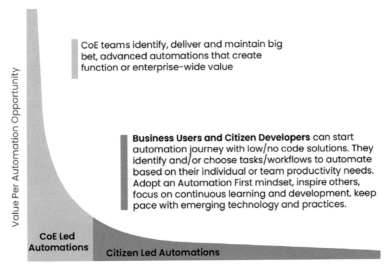

FIGURE 6.10   Automation Opportunities Graph.

in designing this model as citizen-development adoption warrants a lighter touch and governance model.

- **Adoption of standards, policies, and controls**: It is extremely important to establish the right standards, policies, and controls to promote the adoption of appropriate business practices, development, and execution standards and avoid any mistakes in leveraging automation.

Federated/Distributed Delivery Model

In addition to the "automation democratization or citizen-led development" model, we also need to consider the "federated/distributed automation delivery" model to achieve the desired scale and acceleration for adopting automation.

The CoE structure proposes a "centralized model" for establishing, developing, and leveraging a set of core competencies, services, and activities to achieve scale, standards, and benefits of automation within the organization. See Figure 6.11 for a representation of the centralized model. However, it is also important to consider the constraints of a centralized model and factor in any adjustments in the operating model. The two most common constraints that need to consider are (a) resource capacity to deliver on the automation demand and (b) distance from the functional/process knowledge.

FIGURE 6.11   Intelligent Automation CoE Centralized Model.

We need to consider establishing a "federated/distributed automation delivery" model to address both of these constraints for scale and acceleration. This model allows for taking advantage of the CoE structure as well as the distributed delivery model governed by the overall operating framework. However, additional factors such as organizational culture, operating model, financial governance, etc., are to be considered for success.

With this in mind, an appropriate design and alignment to adopt a federated/distributed model are critical for success. See Table 6.8 for an example of how an accountability and collaboration model can be defined between the CoE (center) and the functional (edge) teams to maximize value creation through automation delivery.

Operating Governance: Roles and Responsibilities

In addition to the segregation and sharing or responsibilities of automation development and delivery between intelligent automation CoE (center) and functional (edge) delivery teams (described above in the federated/distributed delivery model), the roles and responsibilities between the business units and the intelligent automation CoE also need to be clearly defined and aligned.

TABLE 6.8   Federated/Distributed Model

Automation Capabilities and Focus	CoE (Center)	Functional (Edge)
1. **Automation ecosystem platform** (architecture, standards, infrastructure, investment, as well as identify, explore, evaluate, adopt and integrate cognitive capabilities such as AI/ML, NLP/NLU, chatbot, and OCR to drive advanced and intelligent automation @ scale)	Own	Adopt
2. **Data, security, compliance, and controls** (drive consistent adoption of cyber security, data and enterprise architecture, business and audit controls standards and policies)	Own	Adopt
3. **Automation ideation, prioritization, and portfolio planning** (identity, explore, evaluate, prioritize automation ideas/opportunities and drive portfolio planning to achieve the intended outcome)	Contribute	Own
4. **Intelligent automation delivery** (design, develop, and deliver automation solutions to enable the touchless operation and intelligent digital workforce)	Own and deliver (shared)	Own and deliver (shared)
5. **Automation democratization** (citizen-led automation, BOT-a-Thon, automation community engagement, training, and enablement)	Own	Leverage and adopt
6. **Governance, value measurement, adoption/usage tracking, and monitoring** (automation portfolio governance, management, value measurement, adoption/usage tracking, and monitoring)	Define and manage	Collaborate and adopt
7. **Support and operations** (ongoing operations, support, maintenance, and data and security policy/controls adherence)	Define and manage	Collaborate and adopt

Business units retain accountability and responsibility for automation identification, prioritization, and benefits realization. At the same time, the intelligent automation CoE operates as a delivery 'factory' and/or performs an independent portfolio management role to track realized benefits. As a result, we see the internal customer and intelligent automation CoE jointly responsible and accountable for benefits in some more innovative business models. The real advantage of this is that it forces the CoE to have skin in the game, so when it comes to processing selection, they only sign up for automation projects where they believe in the benefits.

Table 6.9 presents the different roles, representations, and key focuses across the enterprise for driving automation adoption and value creation.

TABLE 6.9 Role, Representation, and Key Focus Across the Enterprise

Role	Represented by	Key Focus
Automation Operating Committee	Senior leadership within the company	• Align automation capabilities as a strategy to achieve digital transformation vision; make automation a strategic priority • Guide and govern automation value targets and realization • Drive and visible sponsorship of the cultural shift • Approve investment needed
Automation Champions Group (Line of Business)	Various business functions include finance, Legal, Human Resources, Sales, Marketing, Operations, Supply Chain, Customer Support, etc.	• Drive execution-level planning, decision-making, and alignment • Identify, prioritize, and champion automation initiatives aligned to functional strategies and outcomes • Own (be accountable for), manage, and measure business outcomes through automation • Be the change agent, enable the team
Advisory Teams	Various advisory functions: • Information Technology • Enterprise Architecture • Cybersecurity • Internal Audit • Human Resources	• Represent and advise on key considerations for technology strategy, architecture, security, controls, and workforce management in shaping the strategy and design for an intelligent automation ecosystem and solutions • Support developing standards, policies, processes, and controls for these respective areas to drive enterprise adoption and adherence to guidelines
Intelligent Automation CoE (Center)	Centralized team to drive focus areas below to drive enterprise adoption of intelligent automation: • Architecture, Standards, and Governance	• Architecture, Standards, and Governance • Define the architecture, standards, and governance for the integrated intelligent automation ecosystem (includes RPA, AI/ML capabilities, and various adjacent emerging technologies) • Establish the infrastructure architecture design for the intelligent automation ecosystem • Partner with the advisory teams to develop and drive the adoption of standards, policies, processes, and controls for technology strategy, architecture, security, controls, and workforce management • Community and Citizen Development • Drive the focus to build, engage and mature the automation community to promote the democratization of automation through citizen-led automation programs *(Continued)*

TABLE 6.9 (Continued) Role, Representation, and Key Focus Across the Enterprise

Role	Represented by	Key Focus
	• Community and Citizen Development • Automation Engineering and Operations	• Define, execute, and measure the success of various community programs • Evangelize automation technologies, promote automation adoption, champion and support citizen developers • Define and execute the change, communication, and engagement strategy and approaches for different stakeholder groups • Automation Engineering and Operations • Manage the automation engineering and operations focus on delivering the automation capability roadmap • Design, develop, test, and deploy the automation solutions (including integrating automation solutions to the enterprise applications in the most scalable and effective way) leveraging the automation ecosystem platform capabilities (including RPA, AI/ML capabilities, and various adjacent emerging technologies) • Supervise and manage the digital workforce's administration, operation, and controls, executing the business process steps/activities in an automated way. Also responsible for deploying new solutions (digital workers) into the production environment, managing/monitoring utilization, communicating operational highlights to the stakeholders, enabling business continuity planning and ongoing sustenance of the digital workforce
Functional (Edge) CoE	Functional automation delivery teams within any business function outside the centralized Intelligent Automation CoE.	• Work in partnership with the centralized CoE (governed by a "federated/distributed automation delivery" operating model) to develop and deliver functional automation solutions • Adopt architecture, standards, infrastructure, platform, and policies for automation ecosystems and focus on functional automation solution creation and delivery
Citizen Developers	Citizen developers (trained and certified in automation ecosystem technologies) focus on individual or team-level process activities/tasks automation.	• Identify personal or team-level automation ideas • Develop and use automation solutions that drive personal or team productivity improvements • Engage in automation community events and share automations developed with others in the company

Operating Governance: Prioritization and Decision-Making

Establishing a well-defined prioritization and decision-making framework is critical to driving the effective adoption of automation capabilities and value realization from intelligent automation investments.

The decision-making framework must account for agility, the nature of the decision, and the level of impact in determining who should be involved in the decision. See Table 6.10 for a generic decision-making model.

Across the enterprise, we must implement an objective and value-driven automation opportunity prioritization framework. This framework must align with digital transformation priorities and different value drivers for automation in prioritizing focus and opportunities.

To identify and prioritize the right automation opportunities aligned to the digital transformation strategies that drive value delivery through

TABLE 6.10   Generic Decision-Making Model

Nature/Type of Decisions	Frequency	Level of Impact	Accountable Party	Involved/Influencing Party
1. Technology, architecture, and standards	Medium (quarterly)	Strategic	Intelligent Automation CoE	• Information Technology • Enterprise Architecture • Automation Operating Committee
2. Investment and value target	Low (annual)	Strategic	Automation Operating Committee	• Intelligent Automation CoE • Automation Champions Group (Line of Business) • Information Technology
3. Strategic priorities/ top-down objectives	Medium (half-yearly)	Strategic	Automation Operating Committee	• Intelligent Automation CoE • Automation Champions Group (Line of Business)
4. Automation ideation, prioritization, and portfolio planning	High (monthly, weekly)	Operational	Intelligent Automation CoE	Automation Champions Group (Line of Business)
5. Resourcing, capacity planning, methodology, execution alignment	High (monthly, weekly)	Operational	Intelligent Automation CoE	Automation Champions Group (Line of Business)

intelligent automation adoption, we must adopt the right combination of top-down (process heatmap driven) and bottoms (citizen-led ideation) approaches.

In addition, the prioritization framework must apply a robust set of value drivers to maximize value delivery through automation priorities. For example, identifying and prioritizing automation opportunities will be based on each opportunity's quantitative/qualitative outcome.

- Revenue/margin increase
- Cost savings/avoidance
- Customer/partner experience
- Employee experience
- Operational efficiency improvement
- Quality/compliance improvement

Process Heatmap-Driven Opportunity Prioritization
- Establish a common process framework to define Level 1, Level 2, and Level 3 process areas – utilize an already existing framework used within the company or leverage an industry-standard framework.

- Categorize each of the areas of the process map based on the current automation level and impact (which can be measured using the value categories described above) – the current level of automation enables visibility of the opportunity areas. In addition, the impact of automation enables the prioritization of these opportunities.

- Identify and prioritize automation opportunities based on the process heatmap and associate these with the digital transformation strategies and priorities for tracking and measurement.

Citizen-Led Ideation and Prioritization
- Enable framework, process, and tools for employees to identify and submit automation opportunities. Allow for a predefined set of questions to guide the employees through capturing the process characteristics and value-related metrics.

- Establish a collaboration and review process for the automation ideas submitted by the process SME, process owner, and functional leadership to mature the idea as well as value potential.

- Prioritize the automation ideas based on value potential and execution ability (functional CoEs, citizen developers can provide additional delivery capacity).

Financial Planning and Funding

It is important to establish the financial model to support the investment required for scaling the adoption of automation across the enterprise. In addition, the financial model must align with the organization's investment prioritization and approval process requirements.

The investment requirement must be comprehensive to account for different cost categories:

Software Licensing/Subscription

The intelligent automation ecosystem always includes various software components to deliver overall automation solutions. These software components can be from the same product vendor. However, most typically, the automation ecosystem requires products from multiple vendors.

In addition, the software licensing models can be different for different components and/or different vendors. The licensing model is usually subscription-based (annual recurring cost) or consumption-based (the more you consume, the more you pay). Also, the investment model needs to incorporate the implication of contracting models offered/negotiated.

Software licensing/subscription cost must incorporate assumptions for estimated development resources outside the intelligent automation (center) CoE, such as development capacity in the functional (edge) CoE and citizen developers. In addition, if automation is deployed to end-users (usually called attended automation), other licensing/subscription costs may need to be incorporated.

For instance, cybersecurity-required applications (e.g., identity management and PAM), operating systems (on the infrastructure provisioned), project management, training, community engagement platform, etc., must also be considered a part of the overall cost model.

Infrastructure Cost/Consumption

Infrastructure setup and ongoing operating costs are key elements incorporated into the investment framework. Suppose the software vendor provides cloud-based software as a service offering. In that case, the infrastructure cost may already be included in the subscription licensing pricing (the consumption-based pricing may also include an increase in infrastructure cost as the consumption increases). In the cloud-based services model, additional investment may be required to support multiple environments such as product, testing, development, training, etc. If software components require an on-premises implementation, hardware costs, including setup and ongoing operating costs, must be planned and incorporated within the investment model. The hardware/infrastructure cost may differ based on the implementation architecture and design (e.g., private data center vs. public cloud).

Resource Cost/Investment

The resourcing cost/investment pertains to the human resources required to establish the intelligent automation CoE (the specific roles and skills are defined in the sections earlier) and support different activities outlined earlier. In addition, the functional teams may also require additional resources to support the activities related to driving automation adoption within the organization.

The resource investment model must incorporate a baseline cost (core skills and capacity required) and an incremental cost model based on the scaling plan (in relation to the increased volume of automation solution adoption). In addition to automation solution design, development, and delivery, the investment model must include resourcing requirements for ongoing operations of the digital workers (automation solutions).

Program Cost/Investment

Consideration of program investment requirements needs to support the overall automation journey and adoption across the enterprise program. Below are a few of these investment areas that are very critical to plan for in the investment model:

- **Training and certification**: Training and certification of employees in the automation methodologies (design thinking, agile, automation opportunity identification, value measurement, etc.), technologies (RPA, AI/ML, chatbot, process mining, reporting, and analytics, etc.) are critical for successful adoption of automation. The investment model must include cost requirements for developing training/certification programs and the content and delivery.

- **Community engagement**: The automation community (building, engaging, and maturing) is critical for scaling and driving enterprise-wide adoption of automation capabilities within the organization. Investment requirements for implementing a community engagement platform, running community events, rewards and recognition of automation champions, communication and change management activities, etc., must be included in the investment model.

- **Industry event participation**: As the market for automation solutions is rapidly evolving, it is very critical to stay engaged and to participate in industry events such as conferences, user group discussions, and industry forums to aid in planning additional program costs in the overall investment model.

In alignment with the company's investment planning and funding structure, process, and practices, it is important to evaluate and establish the right investment planning and funding (centralized or chargeback). In the case of a centralized investment model, an appropriate approach for investment planning and association to value generation need to be considered, as the value generation accountability may reside with the functional line of businesses.

However, suppose a chargeback model is more appropriate given the organizational context. In that case, additional consideration must be given regarding chargeback models, the complexity of managing, transparency/consistency/accuracy/reporting requirements, cost allocation model, etc.

Value Measurement and Realization

Establishing a value measurement and realization framework is important to ensure that the intelligent automation program delivers the desired business outcome objectives and return on investment. Three aspects of this framework are:

- **Define the value drivers and KPIs**: Define and establish the value categories and key performance indicators to measure the value

- **Identify, capture, and measure the value and operational KPIs**: Processes and methods for identifying, capturing, measuring, validating, and reporting the value and operational KPIs

- **Communicate and share across the enterprise**: The determining process for communicating and sharing the value and operational KPIs

Define the Value Drivers and KPIs

Establish the value drivers for automation and identify the key metrics to measure the value creation through automation. In addition to the value metrics, it is also important to define and measure operational performance metrics for the intelligent automation program.

Value drivers define how value is created through automation. These value drivers must be comprehensive and account for both quantitative and qualitative outcomes.

- **Revenue/margin increase**: Automation of processes to drive (directly or indirectly) revenue and margin improvements. Typically, these processes are related to sales, product introduction, new business models, and/or finance processes dealing with revenue and margin analysis, generation, and recommendation. *KPIs may include* gross sales revenue, average sales price, deal size, discount, gross margin, etc.

- **Cost savings/avoidance**: Automation of processes to reduce manual work and thus require fewer human resources. This is one of the easiest and most common drivers for investing in automation. *KPIs may include* manual hours automated, cost per transaction, headcount reduction or reallocation, cost avoidance as a result of not needing to hire more people or implement expensive solutions, etc.

- **Customer/partner experience improvement**: Automating processes involving customer and partner interactions. Intelligent automation capabilities are leveraged to simplify and automate processes and information sharing to create a better and more personalized experience for customers and partners.

  - This focus can also involve creating automation capabilities for employees to deliver memorable and effective interactions with customers and partners in multiple enterprise areas.

- *KPIs may include* SLA adherence to customer/partner-facing processes, ease of doing business, NPS, etc.

- **Employee experience improvement**: Automation of processes/activities to improve employee productivity and experience. Freeing up human resources from non-value-added repetitive tasks and reskilling human resources to perform higher-value activities allows companies to maximize investment in human capital. In addition, this also enables a company to prepare the workforce for the "future of work" through effective reskilling. KPIs may include productivity, employee engagement/satisfaction, digital literacy/fluency/mastery, etc.

- **Operational efficiency improvement**: Automation of processes to drive operational efficiency improvements such as speed of execution, transactional throughput, etc. KPIs may include a level of automation (automation index) by process areas, turn-around time (TAT)/SLA adherence for processes, cost per transaction, transactional throughput by process, various business operations metrics across functions, etc.

- **Quality/compliance improvement**: Automation of processes to drive higher accuracy, reduced error rate, improved business, and process controls, reduced compliance and audit risk exposures, etc. KPIs may include accuracy/error rates for transactions, number of automated controls, sample size by controls, regulatory compliance exceptions, cost avoidance due to risk exposure, etc.

Identify, Capture and Measure the Value and Operational KPIs

The first steps required are to define the value categories and identify the value metrics and operational KPIs. Then establish the process and methodology to identify these associated with the automation opportunities, capture the details for each automation solution as a part of the design, and measure the same pre- and post-implementation.

During the automation opportunities identification, the details for value and operational KPIs must be captured potentially through a guided questionnaire as a part of the idea submission, qualification, and assessment process. In the case of a top-down automation heatmap-driven approach, the value and operational KPIs become the foundation for identifying areas where automation can transform and deliver value.

It is also very important to apply a consistent methodology for collecting data, adopting standard definitions (e.g., standard metrics, cost per location, and cost per employee), and designing the calculation value and operation KPIs. Therefore, it is best to engage Finance, HR, and Internal Audit teams to collaborate in the definition stage to ensure consistency across the enterprise. In addition, the representation from the Finance, HR, and Internal Audit teams is critical in reviewing the identified value generated and operational improvements from the automation opportunities to validate and potentially certify.

Communicate and Share Across the Enterprise

The value measurement metrics and other operational performance metrics can be prepared, tracked, and communicated in the form of a Balanced Scorecard across the stakeholder groups. This facilitates transparency, collaboration, partnership, and inclusiveness across the organization to mature and scale automation capabilities.

Technology, Infrastructure, and Security

Technology, infrastructure, and security are critical foundational capabilities to drive the organization's scale and adoption of intelligent automation.

To develop the right automation solutions across a varied set of requirements within the enterprise it is best to adopt an "ecosystem architecture concept" with a core technology platform, a set of adjacent, integrated solution components addressing unique capabilities, and a set of peripheral platforms/applications. This enables various supporting processes (for example, ideation platform, community engagement platform, reporting platform, and agile project management platform). The core technology platform is often the RPA platform.

It is important to align the IT infrastructure, policies, and cybersecurity strategies to support the various product vendor offerings.

And finally, similar to the human workforce and associated identity and access management policies, we need to define a process, structure, and standards for identity and access management (authentication and authorization) for digital workers (a.k.a Robots).

The sections below outline the details of these three aspects of technology, infrastructure, and security.

Technology Ecosystem

The technology ecosystem for intelligent automation represents the architecture of the intelligent automation capabilities, including RPA, adjacent technologies such as OCR, AI/ML, NLP, and chatbot, and integration approaches to include other emerging technologies. It also establishes the integration architecture standards and drives adoption of the same within the company (e.g., intelligent automation CoE, functional CoEs, and citizen developers).

In addition to establishing the technology ecosystem architecture, it is very important to set up a standard process to evaluate and incorporate new emerging technologies in the ecosystem architecture to deliver on the automation strategies and to keep pace with the industry innovation. This is particularly important because existing vendors continuously add more capabilities to their automation products/solutions portfolio, and new vendors are coming up with new products/solutions. Therefore, engaging with technology vendors/partners is important to understand the product roadmap closely and incorporate it within the automation ecosystem architecture.

Data are a critical asset for any company. Automation of business processes/activities involves dealing (extracting, processing, analyzing, reporting, and sharing) with data of various types (structured, unstructured, etc.), various sources (enterprise applications, reporting databases, emails, sharepoint, social media, etc.) and various sensitivity levels (PII – personally identifiable information, sensitive information such as compensation, revenue, cost, human resource-related, etc.). Hence, the technology ecosystem architecture must include architectural standards, policies, and guidelines for handling data to automate business processes. See Figure 6.12 for a logical depiction of the "ecosystem architecture concept" with a core technology platform and a set of adjacent, integrated solution components addressing unique capabilities.

Infrastructure and Environment

Infrastructure strategy, standards, architecture, provisioning, and operations are critical elements for enabling the smooth execution of automation solutions and scaling automation usage and adoption across the enterprise. The following aspects need to be considered in designing the infrastructure architecture and standards:

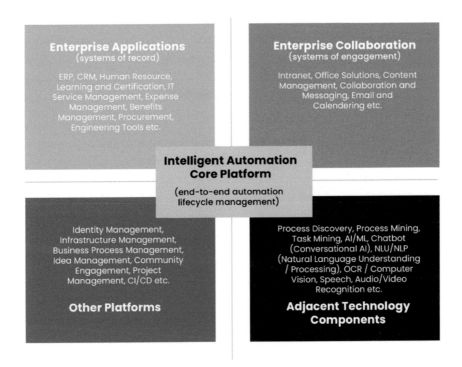

FIGURE 6.12   Ecosystem Architecture Concept.

- **Enterprise architecture and IT infrastructure strategy and standards**: Company's internal IT infrastructure standards and enterprise architecture may require setup and/or implementation of automation within the on-premises data center, private cloud, certain specific public cloud, or automation platform vendor's cloud environment for all or specific processes areas (e.g., internal IT infrastructure standards may require either AWS or Azure or Google cloud infrastructure as standard).

- **Automation platform vendor's services and capabilities**: The automation platform vendor may only have capabilities to support on-premises implementation or the vendor's cloud environment with a software-as-a-service model, depending on customers' needs and standards.

- **Cybersecurity requirements and policies**: Any security policies requiring certain types of infrastructure setup (e.g., healthcare, federal, defense, and financial industries may require on-premises infrastructure for automation for all or some selective processes).

- **Regulatory or compliance requirements**: Any governmental, privacy, or other geo/political policies that require the automation platform/digital workforce to be operating locally or globally (e.g., regulatory requirements for China or other European countries may require local setup for the digital workforce and hence a specific type of infrastructure requirements).

- **Operational requirements**: Any critical requirements for provisioning infrastructure (pre-assigned or on-demand), monitoring, and operational requirements such as high availability, disaster recovery, load balancing, and scaling on-demand. In addition, the operational requirements around multiple environments for development, testing, staging, and production must be considered in the infrastructure design and architecture.

- **Infrastructure cost/investment**: Any investment model requirements for fixed asset depreciation or operational expense with cloud infrastructure, centralized investment or usage-based cross-charging or allocation, etc.

Identity and Access Management

Identity and access management strategies and standards for the digital workforce (a.k.a. robots) are critical for developing and leveraging automation solutions that comply with the specific policies for:

- **Authentication and authorization**: Right to access (authentication) and access to the right things (authorization).

- **Segregation of duties**: Appropriate business and process controls are in place for the digital workers (governed by appropriate authorization). For example, the same digital worker cannot park and post the same journal entry or submit a purchase order and approve it simultaneously.

- **Data access and privacy policies**: Digital workers' identity and access management must be aligned to the stated data access and privacy policies to ensure that the automation solutions are designed and developed with data access and privacy requirements in mind. This also applies to the automation developers and/or automation operations and support team members.

As defined earlier, the "digital workforce" represents the collection of robots (using RPA, chatbots, AI/ML algorithms, no-code/low-code technologies) known as digital workers and the integration of these digital workers in the company's workforce strategy.

When a human worker is hired, they are assigned an Employee ID and User ID with certain authentication and authorization privileges aligned to the human workers' roles and responsibilities. Similarly, when a digital worker is developed and provisioned, assigning a User ID to the digital worker with the appropriate authentication and authorization privileges required for the process/activities is best. This ensures that the automation solution (digital worker) complies with the appropriate policies described above.

Since the User IDs assigned to the digital workers must be administered, it is important to integrate best practices to support the automation solutions with a PAM solution. The PAM tools automate and control the entire process of granting access and passwords to privileged accounts and enable the automation solutions to dynamically and securely access User ID and required authentication without any human intervention.

Engineering and Delivery

Well-designed engineering and a delivery approach are critical to delivering automation solutions at scale. In addition, the approach needs to incorporate various aspects of the automation engineering process to enable standardization, repeatability, quality, and scale.

The overall engineering and delivery framework needs to incorporate a deliberate strategy and structured approach for automation opportunity identification, prioritization, selection, and solution delivery criteria.

Historically, the RPA focus has driven the adoption of the solution and value creation through the "task automation" philosophy. The technology capability enabled and promoted "automate manual, repetitive and mundane activities performed by a human resource" to improve employee productivity, speed, and accuracy. However, the evolution of the technology capability and ecosystem (process discovery, process/task mining, long-running processes, human–robot interaction, OCR, AI/ML integration, end-to-end process automation orchestration, etc.) enabled the possibility of expanding the scope and value generation through automation of end-to-end cross-functional processes.

End-to-End Process/Task Automation

The automation engineering and delivery framework must define a clear strategy and structured approach for a specific process, methodology, expectation, and strategy to deliver solutions based on the automation continuum from individual task automation to end-to-end cross-functional process automation.

The automation continuum spans from basic task automation for an individual employee to sub-process (team level or functional processes) to end-to-end cross-functional processes. See Figure 6.13 for an example automation continuum. In addition, this continuum's engineering and delivery ownership need to be aligned for citizen-led, function (edge) led, and/or CoE (center) led activities.

Each of the frontiers of automation (task automation/citizen led, sub-process automation/function, or CoE led and end-to-end process automation/CoE led) has a very unique style and requires a unique execution approach. A clear strategy across these frontiers of automation should guide the approach for speed, value, volume, quality, and governance considerations to manage the stages of the automation life cycle (ideation

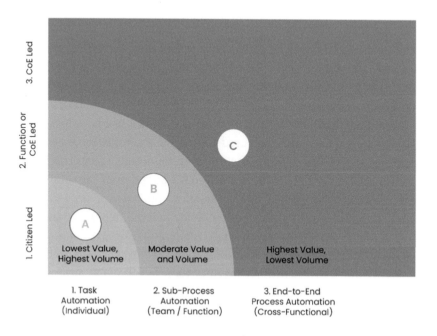

FIGURE 6.13 Automation Continuum.

through delivery). This will maximize the overall impact of the automation program at the enterprise level.

Table 6.11 provides a list of engineering and delivery processes. Also, refer to sections earlier in the book for details on process heatmap-driven and citizen-led automation ideation and prioritization.

Automation (Apps/RPA) Approach

The automation engineering and delivery framework must incorporate a structured approach to ensure a close alignment with IT strategy and plan for

TABLE 6.11   Engineering and Delivery Process

Engineering and Delivery	Task Automation	Sub-Process Automation	End-to-End Process Automation
1. Identification	Crowdsourcing (employee submitted) ideas. Use of an ideation portal with structured and guided idea submission.	Function/team-based workshop, process walk-through, or process performance metrics-driven. Use process mining or task mining tools and an ideation portal.	Process heatmap and top-down value target driven. Use of process mining or task mining tools.
2. Prioritization and planning	Value criteria (typically lower in the spectrum) and capacity of citizen developers. Planning is done at an individual level (based on time commitment and capacity of citizen development).	Primarily value criteria (expanded dimensions) and functional goals. Planning is done at the Team or function level.	Alignment to digital transformation strategy and objectives, top-down value targets. Planning is done at the enterprise level with the Automation CoE.
3. Automation engineering	Individual citizen developers. Heavy usage of standards, reusable components, etc. The only usage of pre-approved automation capabilities, integrations, etc.	Functional (edge) development team. May involve multiple automation capabilities and requires some level of CoE.	Enterprise automation CoE team will involve many automation capabilities and typically custom integrations with many enterprise applications and data sources.

(Continued)

TABLE 6.11 (*Continued*)   Engineering and Delivery Process

Engineering and Delivery	Task Automation	Sub-Process Automation	End-to-End Process Automation
4. Delivery and deployment	Through the enterprise automation CoE deployment process (automated for the most part)	Through the enterprise automation CoE deployment process	Due to heavy dependency on other IT programs, the enterprise automation CoE deployment process will need to coordinate with the IT release and deployment plan.
5. Operations and sustenance	Individual citizen developer	Functional (edge) team	Enterprise automation CoE team
6. Quality and governance	Standards, policies, automation platform architecture, automated engineering, and quality control processes. The only usage of pre-approved automation capabilities, integrations, etc.	Standards, policies, automation platform architecture, automated engineering, and quality control processes. Selective review by CoE and advisory teams (IT Enterprise Architecture, Cybersecurity, etc.)	Standards, policies, automation platform architecture, automated engineering, and quality control processes. Review by and involvement of advisory teams (IT Enterprise architecture, Cybersecurity, Internal Audit, Human Resources, Finance).
7. Value realization	Individual productivity. Self-reported and automation solution telemetry data.	Team productivity, process throughput, process quality, process SLA, cost savings/avoidance, etc. Process performance measurement (through the process or task mining).	Customer/partner experience, employee experience, revenue/margin increase, cost savings/avoidance, etc. Process performance measurement (through the process or task mining), finance validated revenue/margin/cost impact.

enterprise architecture, infrastructure standards, cybersecurity requirements/policies, and operational monitoring/controls/management. This ensures that "automation solutions are designed and developed right" from the start.

In addition, it is also very important to establish a structured approach to determine the "right automation solution" for a specific process/task automation opportunity. The automation philosophy (as touted by the vendors and practitioners) for robotic and intelligent process automation can be interpreted as engineering and delivering these automation solutions to address challenges and gaps in the IT department's capabilities.

The intelligent automation or robotic process automation approach and capability enable developing automations with significant value, speed, and scale in addressing a genuine gap of opportunities, priorities, capabilities, and speed in the "typical centralized IT-driven approach." However, engaging and integrating this framework with IT is paramount to adopting intelligent automation capabilities successfully.

One such area is determining the right automation solution for a specific process/task automation opportunity. This approach of determining if RPA or intelligent automation solution is the right interim or long-term automation solution (enabling alignment, timing, and value expectation) is essential to maximize the value potential and minimize the creation of any technical debt for the enterprise. See Figure 6.14 for a decision tree related to implementing an automation structure.

Methodology, Process, and Tools

The automation engineering and delivery methodology must incorporate a standardized approach for the automation solutions' entire life cycle (idea-to-outcome). In addition, the methodology, processes, and tools must be designed to enable agility, scale, and the right level of control within the automation operating model.

Within the overall automation operating model, the life cycle of automation can be largely broken down into three primary stages: (a) Ideation and Identification, (b) Engineering and Delivery, and (c) Operations and Sustenance. Each stage will require a specific approach, engagement, structure, and tools integrated through automation and process. See Table 6.12 for the lifecycle of automation.

The CoE must create a playbook incorporating all the stages of the automation lifecycle. The playbook defines various aspects of the operating model tailored to suit the organization's structure and culture, existing methodologies/approaches, and digital transformation agenda.

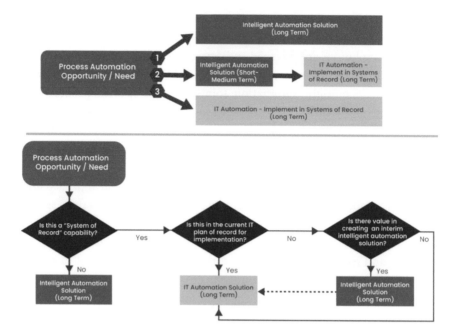

FIGURE 6.14 Decision Tree Related to Implementing an Automation Structure.

Operations and Sustenance

A well-structured and enabled strategy to operate and maintain the "digital workforce" (a.k.a. BOTs) is a critical component to scale the adoption and stability of Intelligent Automation capabilities. The right focus, skills, and model for operations and sustenance of the "digital workforces" ensure the realization of intended business value throughout the end-to-end lifecycle of intelligent automation solutions.

While establishing an operating model for intelligent automation adoption, a lot of focus is usually on the upfront process of awareness, engagement, ideation, prioritization, engineering, delivery, and value realization. However, one of the most critical aspects of sustained value realization and scaled adoption is establishing the operations and sustenance structure with the right resourcing, process, tools, and measurements. These aspects include the following:

- BOT operations and monitoring
- Process SLA and execution accuracy
- BOT performance and capacity

TABLE 6.12 Life Cycle of Automation

Ideation and Identification	Engineering and Delivery	Operations and Sustenance
The ideation and identification stage focuses on activities and approaches to ideate and identify automation opportunities. These activities must account for both a top-down strategy and target-driven approach and a bottoms-up crowd-sourced ideation approach.	The engineering and delivery stage focuses on prioritization and planning, automation solution engineering, and delivery of the solution into operation. The automation solution may involve multiple automation capabilities and various enterprise applications. Hence, it must incorporate appropriate design and alignment for security, privacy, workforce policy, architecture standards, etc.	The operations and sustenance stage focuses on operating the digital workforce (BOTs) and ongoing maintenance. This stage focuses on ensuring that the digital workforce (BOTs) are operating as expected and executing the automation as per defined expectations (SLA, Throughput, etc.), managing and recovering from any service failures, and ongoing maintenance of the digital workforce in light of any environmental changes.
The methods and activities may include awareness events, ideation campaigns/competitions, functional workshops, process heatmaps, and automation maturity analyses.	The methods and activities must incorporate specific variations for three distinct engineering styles: citizen development, function (edge) development, and CoE (center) development. The activities in this stage include prioritization (backlog), planning, design, development, testing, and deployment. The engineering and delivery process must be designed based on the Agile Framework/ Methodology and adopt the Agile Manifesto and Principles. If the organization has already adopted agile methodology and tailored it to suit the culture, it is best to adopt it.	The methods and activities must incorporate variations due to operating models of 3 different styles of digital workforce engineering: citizen-developed, function (edge) developed, and CoE (center) developed. In addition, the operations and sustenance must incorporate processes related to 24×7 support, SLA management, capacity scaling, business continuity planning, and invocation,
Various tools involved in driving the activities for ideation and identification are: always on the ideation platform, process mining, task mining, process heatmap, automation maturity analysis tools, enterprise architecture, etc.	Various tools involved in driving activities within the engineering and delivery stages are the Agile Framework/ Methodology, project management tools, automated testing tools, code repository/version control, automated release pipeline, deployment management, etc.	Various tools involved in driving the activities for operations and sustenance are support requests and ticketing, digital workforce (BOTs) monitoring and operations, etc.

- Business continuity
- Incident management
- Support request management
- Issue resolution process
- Ongoing maintenance
- Enhancement management
- Automation ecosystem upgrade and maintenance
- BOT retirement and decommissioning

BOT Operations and Monitoring

The BOTs (Digital Workforce) deployed through the engineering and delivery process will need to be operated to execute the process activities and monitor the successful execution of the designed automation steps and the intended outcome (Manchester & Cope, 2019). This focus on operations and monitoring ensures that digital workers perform the activities within the defined SLA, processing the transactions as designed and dealing with any fallout in a timely manner.

With the discussions on the digital workforce strategy and integration of the "digital worker" in the organization structure (how we can deliver the work/outcome through an optimal mix of human and digital resources), we need to design and adopt an ownership model for digital worker operations and monitoring accordingly. See Table 6.13 for the ownership model.

Furthermore, this ownership model must also incorporate implications of three different styles of digital workforce engineering: citizen-developed, function (edge) developed, and CoE (center) developed.

The BOT operations and monitoring processes need to manage the following:

- BOT execution and scheduling
  - The process automation design must define the expected invocation method (on-demand, scheduled, event-triggered, etc.) for BOTs. The invocation method may vary depending on the business criticality, operations time, geography, and other parameters specific to the process. For example, automation for the Finance

TABLE 6.13 Ownership Model

Centralized Ownership	Decentralized Ownership	Shared Ownership
The Automation CoE owns all digital workers' operations and monitoring responsibilities (BOTs) in the centralized BOT operations and monitoring model. Therefore, the functional line of business managers responsible for the specific process SLA, performance, and outcome relies on the Automation CoE for understanding how the BOTs are performing and typically submits support requests for any changes. In this model, citizen developers will own operations and monitor individual task automation solutions within the standard policy and framework. This may be a good starting point for many CoEs due to the need to develop skills, experience, and many CoEs due to needing to develop skills, experience, processes, and tools. However, this is also one of the most constraining aspects of scaling automation adoption and the three unique styles of automation engineering. In addition, this model represents a very traditional IT automation paradigm and does not promote the "digital workforce strategy" outlined in the chapter earlier.	The functional line of business managers owns digital workers' operations and monitoring responsibilities (BOTs) pertaining to the process ownership model. Essentially, the digital workers (BOTs) become part of the organization structure and resource mix (human and digital) within the business line managers' function/group/team. The Automation CoE assumes ownership of enablement, tooling, and infrastructure-level operations and monitoring. The business line managers assume full accountability for delivering the work through the optimal human and digital workforce mix. This model aligns better with the "digital workforce strategy" outlined earlier in the chapter. The business line managers manage running (scheduling, starting, and stopping) the BOTs within their scope of responsibilities and monitor the performance of the BOTs along with fallout handling. The citizen developers will own operations and monitor the individual-level task automation solutions within the standard policy and framework. While this accelerates the scaling of automation adoption and aligns with the "digital workforce strategy," this model requires enabling and adopting specific tools, processes, and controls to accommodate the enterprise-level complexities.	This model enables shared accountability and responsibilities for operations and monitors all digital workers (BOTs). Moreover, the ownership can be established based on the automation style and scope: Automation (center) CoE: • Enablement, tooling, and infrastructure-level operations and monitoring • End-to-end cross-functional process automation solutions Functional Business Line Managers: • Functional or team-level process automation solutions Citizen Developers: • Individual-level task automation solutions (governed by standard policy and framework) This model allows for the most flexibility and scale for automation adoption within the enterprise.

Process may vary in the execution frequency during normal operating days, month-end, quarter-end, and year-end days.

- The BOT operations team ensures that the BOTs operate per the predefined invocation method.

- The scheduling of BOTs needs to incorporate any interdependency of the automated processes.

- BOT accuracy and fallout

  - The accuracy of and fallout from the automation solutions must be defined, reported, and monitored as a part of operations and monitoring activities.

  - The handling requirements, process, and ownership for fallouts and exceptions must be agreed on and well-defined. For example:

    – If certain types of transactions are not handled or excluded as a part of exception handling.

    – In the case of document processing (OCR solutions), there may be a confidence threshold below which human intervention is required.

- BOT performance and SLA

  - Monitoring BOT performance and adherence to predefined process SLA and throughput must be in place to maintain expected business performance.

  - The SLA and throughput requirements for processes should be defined at the design time, monitored, and reported as a part of operations. The SLA and throughput requirements may vary depending on the business criticality, operations time, geography, and other parameters specific to the process. For example, the SLA and throughput expectation for automation for Finance Process may vary in execution frequency during normal operating days, month-end, quarter-end, and year-end days.

  - Both BOT capacity and scaling methods must be defined, planned, and executed to meet the SLA and throughput requirements. This scaling may involve provisioning more BOTs (digital workers) and division/balancing of workload appropriately among the available BOTs. In addition, the automation platform and infrastructure may allow for "auto-scaling" capabilities whereby provisioning and workload balancing can be automated based on predefined performance thresholds.

- Business continuity planning and invocation

  - Business continuity planning and invocation ensure that business operations can continue according to predefined policies and expectations when operational disruptions occur with the digital workforce.

  - Digital workforce business continuity planning is typically part of an overall business continuity planning framework defined and adopted within the enterprise.

  - It is important to establish the requirements for business continuity, ownership, and criteria for invocation at the time of automation design. In addition, the functional line of business teams may need to account for appropriate resource planning depending on the business continuity plan.

Incident Management

The operations and sustenance model includes establishing a structure, process, resourcing, and tools for incident management to support the digital workforce. The incident management process must address the following two critical aspects of digital workforce operations.

- **Support request management**: defines how end users report issues and submit questions about the automation solutions

  - If there is an enterprise-level support ticket management process and tool, it is best to adopt the same with adjustments to tailor to the need for intelligent automation solutions. This is critical to driving adoption (as end users are already familiar with the existing ticketing process and tools) and yet accounts for the uniqueness of intelligent automation solutions.

  - The support request management process must include a definition of incident submission (manual and automated), assignment, and collaboration between teams to address the incident.

In addition, support incidents can be submitted by end-users and digital workers (BOTs) as an exception-handling mechanism.

- In the case of automated support incident creation as part of exception handling by digital workers (BOTs), it is important to establish standards for issue categorization, severity, priority, association to process/automation solutions, issue description, and assignment rules.

- **Issue resolution**: defines how issues are identified, triaged, and resolved, along with root cause analysis and preventive action planning.

  - The issue resolution process includes process, reporting, and communication around identifying an issue with the BOT's operation, performance, and/or accuracy, triaging the issue with other teams, driving resolution of the issues, performing root cause analysis, and implementing preventive actions.

  - The issue can be triggered manually or through automated monitoring by the overall ecosystem and changes within the environment.

  - It is important to understand the nature of the issue, its severity, and its impact on overall business performance. The severity and impact of the issue may vary depending on the business process (financial, customer-facing, etc.) being impacted, the time of operations (month-end, quarter-end, etc.), and the potential cause of the issue (process, automation solution, infrastructure, etc.).

  - The issue resolution process must define a clear collaboration and engagement protocol involving various functional teams, application teams, infrastructure teams, and vendor support teams such as automation platforms/infrastructure providers). This ensures a faster resolution process and enables clear communication and sharing.

  - The issue resolution process for automation solutions must integrate into overall incident/crisis management, disaster recovery planning, and business continuity planning frameworks and protocols.

- The process must also include a root cause analysis and retrospective and preventive action planning to build a knowledge base of issue resolution and implement corrective actions.

Ongoing Maintenance

The intelligent automation ecosystem and automation solutions require ongoing upkeep/maintenance to ensure appropriate stability and resiliency for the solutions. Therefore, it is important to continuously adjust and account for environmental changes for automation capabilities and solutions. The environmental changes could come from business process changes, enterprise application upgrades/changes, infrastructure upgrades/changes, vendor-driven automation capabilities upgrades/changes, and minor enhancements for the digital workforce.

Enhancement Management

- Enhancement management is a key part of driving robust operations of digital workers to deal with changes in business processes, enterprise applications, and new automation requirements. This ensures that digital workers (BOTs) continue to deliver business value and do not become obsolete.

- The enhancement management process for digital workers must be integrated into the automation engineering and delivery methodology and processes, even though the operations and sustenance teams can implement some enhancements (typically minor).

- This process also must incorporate variations to account for three styles of automation engineering: citizen-developed, function (edge) developed, and CoE (center) developed.

Automation Ecosystem Upgrade and Maintenance

- One of the critical activities owned and driven by the operations and sustenance team is maintaining the automation ecosystem environment up to date with the patches and upgrades for the infrastructure, vendor solutions/automation platforms, and upgrades of the enterprise applications.

- Usually, the patches and upgrades to infrastructure, automation platforms, and enterprise applications are intended to either fix existing issues and/or release/deploy new capabilities/functionalities.

- The upgrade and maintenance process must include an appropriate level of activities for applying the patches or upgrades, performing regression testing and validation, and deploying/enablement the patch or upgrade in the production environment in a coordinated fashion.

- This also involves appropriate levels of communication across all the stakeholder groups, allowing transparency and collaboration in keeping the automation ecosystem up to date.

BOT Retirement and Decommissioning

- In addition to developing new automation solutions and driving the adoption of automation capabilities within the enterprise, it is also important to establish a structure and process for retirement and to decommission automation solutions when no longer needed.

- In line with the changing business environment and landscape, business processes, enterprise applications, and automation solutions may also become obsolete.

- Robust and ongoing monitoring (and engagement with the business stakeholders and IT stakeholders) may highlight that an automation solution may no longer be needed or not deliver the intended business value. Therefore, appropriate actions must be taken to retire and decommission the automation solution in such a scenario.

- The retirement and decommissioning process must be comprehensive to account for the following:

  - Software license/subscription decommissioning

  - Hardware/infrastructure decommissioning

  - Disabling/de-provisioning of digital IDs for the digital workers

  - Archiving of historical data for any audit/legal purposes

  - Communication with all stakeholders

Cultural Transformation

The focus of cultural transformation in this chapter is based on assessing the awareness and readiness of the employees within an organization with an emphasis on leader humility that invigorates awareness and readiness via the proactive behaviors of employees (Chen et al., 2021). Therefore, some of the warning signs that there may not be alignment between humble leaders and proactive employees include: senior leaders not offering training programs such as bot-a-thons; managers not seeking feedback from employees about the changes; managers not showing appreciation for the contributions of employees including rewarding and sharing success stories; managers not encouraging the innovative, creative ideas from employees; employees not recommending ideas to reengineer more efficient tasks; employees not recommending innovative ideas; employees not being participatory in the employee development process.

Cultural transformation, noted herein, focuses on the human culture approach to change management, including the following three dimensions: employee development, mistake tolerance, and accurate awareness. In addition, creating a forward-thinking action plan is important to mitigate any disconnect between humble leaders and proactive employees.

An action plan can accomplish this.

A digital workforce action plan is required to mitigate the chances of a disconnect between humility leaders and proactive employees. The action plan for preparing the digital workforce includes a shift in the technical tools and techniques and the communication and readiness of human and non-human resources, including considerations related to **culture, processes, people, procedures, and operations** (Project Management Institute, 2017). This cultural transformation was introduced at the beginning of the chapter but has been expanded to include a continued action plan to support cultural transformation, as noted in Table 6.14, column 5. This table includes the action plan criteria/dimension, dimension of humility, what to do, what to consider, and how to continue to support cultural transformation.

Organizations can measure their cultural transformation by conducting the following:

- **Assessing employee attitudes and behaviors**: Are employee attitudes and behaviors changing based on employee survey responses? This will require periodic assessments within a given time frame.

- **Identifying impacts on business performance**: Are key performance indicators being met, such as the number of creative and innovative suggestions presented and implemented? If the targeted number of suggestions has not been implemented, then identify the reasons why and adjust accordingly.

TABLE 6.14   Continual Support of Cultural Transformation

Action Plan Criteria/ Dimension	Humility Dimension	What to Do?	What to Consider?	How to Continue to Support Cultural Transformation?
Culture	Mistake-tolerance	Create a humble culture environment	Honest, creative, and learning environments	Share success stories related to change
Processes	Accurate awareness	Identify changes to the business processes to support the change	Functional or enterprise implementation	Reengineer business processes to remove inefficiencies and bureaucracy
People	Employee development	Identify roles and prepare for the future workforce	Training and retooling	Provide incentives to train and retool (new job role, increase in pay)
Procedures	Employee development	Identify the methodology, tools, and techniques that will be used to implement the automation	Agile and program management methodologies	Conduct bot-a-thons to learn more about automation
Operations	Accurate awareness	Identify technology	Vendors or in-house development	Conduct training to move vendor functions and maintenance in-house to create more awareness and opportunities
Operations	Accurate awareness	Determine infrastructure and security needs to support the automation	Vendors and/or in-house changes	Conduct training to move vendor functions and maintenance in-house to create more awareness and opportunities

- **Tracking and monitoring key milestones**: Are key milestones, such as the number of employees retrained and retooled in intelligent automation tools and techniques, being tracked and monitored? This will include assessing whether the employees have met their commitments related to training and development.

- Continuously identifying impacts, adjusting, and tracking and monitoring key milestones.

CONCLUSION

This chapter included a structural implementation approach focusing on organizational, operational, management, and measurement models and the need for diversity in human and non-human resources used to prepare, govern, and support automations. We conclude with lessons learned from previous implementations and make recommendations related to phased approaches.

CHAPTER 7

Challenges Scaling IA – The Race to IA Takes Off

Neeraj Mathur

VMware

INTRODUCTION

In nearly every industry and market sector, businesses look for innovative ways to reduce costs, increase productivity, improve customer experience, streamline operations, and grow revenue. In the past decade, digital native businesses have disrupted the ways by which they interact with customers, partners, and employees by introducing novel and easily accessible channels of interaction. Forward-looking and data-intensive businesses, especially in finance, healthcare, and technology sectors, have harnessed the benefits of growing capabilities of artificial intelligence (AI) and machine learning (ML) in their business processes. Even traditional businesses have explored and implemented ways to reinvent themselves to remain relevant and competitive by utilizing the power of AI/ML and analytics.

In the last few years, although at a limited scale, intelligent automation (IA) programs across industries exhibited the potential to improve operational cost, employee productivity, customer experience, process efficiency, regulatory compliance, and employee satisfaction. In some organizations, the application of IA has even grown beyond simple rule-based RPA to implement advanced cognitive solutions. Some organizations have made crucial investments in talent to address the skill gaps and change management programs to empower employees to embrace the change. Business leaders have started viewing IA as a technology that can

DOI: 10.1201/9781003276128-7

transform business operations and create value. Digital transformation leaders are more confident than ever about transforming their business processes using IA to achieve higher speed, accuracy, and compliance. All this rise of IA has gone to the extent that IA is now adopted as a critical part of the digital transformation strategy in some organizations.

The broad spectrum of IA technologies enables these organizations to transform business processes whereby a "digital worker" can perform wide-ranging activities from "acting" like a human to "thinking" like a human. IA exploits cognitive and AI technologies, including ML, natural language processing (NLP), and natural language generation (NLG), with robotic process automation (RPA) to build "smart" workflows that can learn and adapt in near real-time. These smart workflows can manage and process structured, semi-structured, and unstructured data that are an integral part of business processes nowadays. More importantly, these digital workers can learn by obtaining human input in case of any exception from the mainstream flow.

Due to the inherent characteristics of the RPA technology, a working bot can be developed in a few days or weeks. On the contrary, it would take weeks or months to build similar capabilities using legacy technologies. This quick win gives the impression, or rather illusion, that the same pace can be maintained while growing and scaling the program enterprise-wide. The organizations that have implemented RPA/IA programs successfully recognize that the program's success is not dependent on technology alone. Rather, it is influenced by many other aspects of the program, including people, governance, and operation. These organizations acknowledge that establishing a strategic and sustainable IA program goes far beyond just developing incrementally more bots to automate additional tasks. It requires the IA program to be a critical part of the digital transformation strategy, it requires a vision and strategy that enables the discovery of innovative automation opportunities, it requires constant engagement with business and IT to develop novel solutions to improve business processes, it requires open and candid communication for cultural transformation, and it requires C-suite sponsorship and funding to sustain the growth (Deloitteeditor, 2020).

Scalable Intelligent Automation Program

The scalable intelligent automation program typically refers to deploying IA across the enterprise to automate cognitive, high-valued, cross-functional, departmental, and inter-departmental processes throughout the organization's business ecosystem. Formally, a scalable IA program encompasses a well-defined strategy augmented by agile governance and operational processes, supported by a collection of IA technologies and run by a group of skilled professionals to accelerate the IA adoption enterprise-wide.

It is important to note that the goal of a scalable IA program is not merely to manage the periodic uptick in the workload and volume of business transactions. Rather, ensuring that a consistent enterprise-wide approach is established to develop, deploy, support, and capture value from IA is much more significant.

Scaling an IA program is an uncharted area with greater complexity and sophistication. Thus, to truly leverage IA to its full potential resulting in enterprise-wide transformational outcomes, it is pragmatic to view the scalability holistically by combining the various aspects from a functional, technical, and governance standpoint. On the functional side, scaling up could mean going from small and low-complexity tasks to highly complex cross-functional or inter-departmental process automation aligned with business objectives and priorities. On the technical front, the scaling up could mean going from one simple digital worker to thousands of digital workers developed by integrating a series of IA technologies that are deployed on a secured and scalable infrastructure. Finally, from the governance point of view, scaling could mean having enterprise-wide secured and agile access to digital workers while encouraging innovation-driven cultural transformation supported by upskilling training and development.

This chapter discusses scaling IA programs from multiple different perspectives. The text begins by developing a deeper understanding of various functional and technical facets to consider while scaling an IA program. Further, it addresses the IA technology outlook, continuous improvement mindset, and holistic visibility of activities needed to establish and manage a scalable IA practice. Finally, the chapter talks about various challenges one might experience or at least should anticipate and provides relevant recommendations about how to remediate them. This chapter also covers some key best practices essential to having a scalable IA program across the enterprise.

FIVE FACETS OF SCALABILITY

At the onset, it is important to recognize that scaling the IA program is unfamiliar territory for many organizations and digital transformation leaders. Moreover, this area brings in novel and unique complexity and challenges not previously experienced by IT or business leaders. Besides, there is no "one-size-fits-all" approach to scaling IA programs, and every organization must go through its unique journey driven by its priorities and policies.

It is also essential to acknowledge that IT and business leaders have different vantage points for the scalability of the IA program. Business leaders consider scalability from the functional viewpoint involving business functions and respective business applications, whereby scaling up could mean onboarding a new business function or business application, whereas IT leaders typically view scalability from the technical architecture of the IA ecosystem and the supporting infrastructure, whereby the scalable IA practice could mean ease of integrating emerging technology and deploying scalable infrastructure to support it.

The IA scalability must be viewed from multiple independent yet related facets to build and manage a scalable IA program across the enterprise. To appreciate the complexity behind achieving enterprise-wide IA and be prepared to handle it, we first need to understand various facets of scaling IA. These facets span across business functions and run through complex business processes supported by many business applications. From business subject matter experts (SMEs) to citizen developers, these facets also shed light on the most critical ingredient for success – people. Furthermore, these facets also address the vital component of IA architecture and infrastructure involving integrating various existing and emerging AI technologies. Please refer to Figure 7.1 for a complete view of all five facets of scalability.

1. **Business functions**: The most common viewpoint to scale the IA program is by progressively introducing IA capabilities across numerous business functions (Finance, Tax, Operations, Customer Service, Supply Chain, Regulatory Compliance, HR, Legal, IT, etc.) of an organization. This facet of scalability focuses on the strategy, approach, and tools required to strategically enable the expansion of the IA program aligned with business objectives, goals, and priorities.

2. **Business applications:** Each business application or system presents unique nuances regarding application access, data security, and management of multiple environments to develop, test, and deploy automation. To avoid a critical roadblock to scaling the IA program, these requirements must be managed proactively. This facet of scalability addresses approaches and techniques to automate an increasing number of business applications.

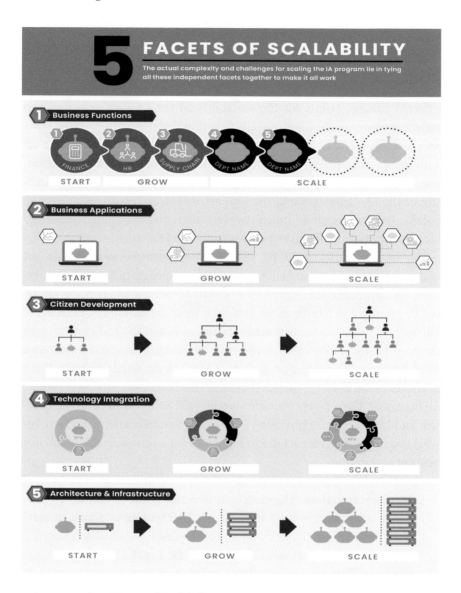

FIGURE 7.1   Five Facets of Scalability

3. **Citizen development:** In the context of scaling the IA program, citizen development is the most common topic of discussion predominantly due to the challenges and benefits organizations experience by scaling across the organization. This facet of scalability involves enabling and empowering business users to become citizen developers to build digital workers for themselves and their departments.

4. **Technology integration:** The fundamental notion of tangible IA is meaningfully integrating existing and emerging IA technologies. These integrations bring efficiency to business processes, improve data-driven decision-making, and truly bring the "intelligence" of the IA into the picture by enabling business processes with AI-enabled technologies. This facet of scalability covers the ever-evolving landscape of IA technologies and their usage to leverage the full potential of IA.

5. **Architecture and infrastructure**: The IA architecture and infrastructure should allow steady growth and agility to build a sustainable and scalable IA program. Technically speaking, the complexity of the IA program lies in building a robust architecture and secured, scalable infrastructure that should facilitate the increasing number of digital workers, the growing number of transactions, and the rising number of hardware nodes. This facet of scalability discusses the IA ecosystem that consists of IA platforms and infrastructure components.

The actual complexity and challenges for scaling the IA program lie in tying all these independent facets together to make it all work. Thus, IA leaders should see beyond the obvious to maintain a complete view across all facets of scalability while focusing on generating value for the organization by leveraging the full potential of IA.

For instance, to develop a sustainable and scalable IA program:

1. IA leaders must play a balancing act between disparate requirements of business functions while maintaining standardized IA technical architecture. For example, for a healthcare organization, there is immense value in having standardized IA architecture to support unstructured data of patients' medical records and unstructured data from their legal and finance departments. Similarly, a manufacturing organization will benefit from standardized architecture to capture and manage Internet of Things (IoT) sensor data and supplier and vendor data.

2. IA leaders must ensure that technology integration practices are aligned with the objectives of each business function to address their specific needs. For example, the call center function will heavily use conversation AI, whereas the finance function might have a more

significant need for intelligent document processing (IDP). These specific functional requirements will require specialized integration with their respective workflows.

3. IA leaders must maintain consistency and agility in onboarding business applications for COE and citizen developers alike. For example, the governance rules and operating procedures should clarify which business application is acceptable for citizen developers, what actions are permissible on these applications, and when to adopt citizen developer practices vs. professionals from CoE.

4. To minimize complexity and inefficiencies, IA leaders must strive to establish standardized policies and procedures for the ever-evolving IA technology landscape. For example, the list of vendors supporting unstructured data is rapidly growing, some of which have specialized capabilities, and others are targeted for relatively general use of platforms. Thus, choosing the right product and vendor aligned with enterprise-wide requirements is increasingly challenging.

5. To establish a successful citizen developer program for each business unit, IA leaders must consider the business function's specific requirements and the skill levels of citizen developers in each department. For example, developing automation involving spreadsheets and email might be sufficient for some business users. In contrast, others might require to be equipped with building an intelligent and fully functional app that can be used through multiple devices like laptops, tablets, and smartphones.

6. IA leaders must establish consistent access management across disparate business applications and functions for digital workers. For example, digital workers across all functions and applications must be granted the least possible privilege required to run business workflow effectively.

7. While leveraging the full potential of IA platforms, IA leaders must also prioritize native automation capabilities available for business applications. For example, many IT, finance, and sales platforms natively support automation, which must be used appropriately. It can produce more effective solutions in many cases than building a new solution on a different IA platform.

In essence, the IA leaders must be able to synergize the competing and disparate activities across each business function and for every IA technology. The modernization of the IA ecosystem by adopting innovative and evolving IA technology is critical for the sustainable growth of the IA program. At the same time, the standardization across business processes, IA ecosystem's components, IT processes and policies, business application access, and CoE governance and operations are essential for the foundation of a scalable IA program. The complexity lies in balancing this modernization and standardization across all five facets of scalability.

Business Functions

Optimizing and automating business processes across enterprise-wide business functions is a powerful way to scale the IA program. There isn't a preferred business function to start with, though the journey of IA typically begins in the finance department owing to numerous manual tasks and the availability of accountants with analytical acumen and proficiency in Microsoft Excel Macros development. To scale the IA program, it doesn't matter where it begins in your organization; it's important how it grows into other business areas.

This facet of scalability addresses the strategy, approach, and tools to enable the expansion of the IA program across the organization. To scale the IA program within a business function or across the enterprise, IA leaders must first address nuances related to each business function. This is required because certain business functions will require special attention to data privacy concerns, exceptional service-level agreements (SLAs) to manage, specialized hardware (IoT, Robotics) or software (legacy) systems, or larger data volumes to manage, and some functions will have an atypical combination of most of the above.

To scale the IA program enterprise-wide, most importantly, IA leaders must identify and define the strategic value of IA both across the enterprise and for an individual business area. The strategic value must align with the organization's objectives, goals, and priorities. IA leaders should also establish the preferred process optimization and automation opportunities for each business area which must be aligned with the established strategic value for each business area. Further, the selected types of process automation will assist in choosing the right combination of IA technologies for the business area. Finally, IA leaders must build a long-term strategic roadmap aligned with business objectives, goals, and priorities to sustain the growth of the IA program (Analyst et al., n.d.).

Identify Strategic Value

Typically, the adoption of IA technologies starts with bottom-up task-level automation driven by a small and siloed proof of concept. Although these automations are essential, organizations must swiftly transition to high-valued end-to-end process automation with necessary sponsorships and adequate funding from senior leadership.

IA leaders must work with senior leadership to recognize and institute the strategic value of the IA program aligned with business goals. The strategic value must be set at the organization and the individual business area levels. IA leaders must identify opportunities to augment business processes by integrating AI-enabled capabilities to unlock long-term strategic business value. Some of the traditional strategic values that organizations have considered are listed below. Each business area should prioritize its combination of these values associated with its business processes.

- Reduce or optimize cost

- Increase revenue growth

- Improve process efficiency

- Improve customer experience

- Reduce regulatory and compliance risks

- Increase employee productivity

- Improve employee satisfaction

- Many more

Identify Types of Process Automation

Vendors and industry publications often promote robotic process automation as a quick solution to reduce headcount, save money and improve customer service. As a result, business partners often pressure software engineering leaders to focus on the tactical needs of 'routine task automation' with RPA. However, not all processes are simple, routine, and short-running. Many processes involve complex business rules, intelligent automated decision making and optimization. Data is always not structured.

These processes and tasks can be categorized as follows, assisting in uncovering the opportunities to automate using the IA ecosystem:

1. **Small tasks** are usually performed by individuals and are limited in scope by the duties of the team they work for. Typically, the automation of these tasks is focused on employee productivity and cost optimization.

2. **Self-service tasks** are characteristically low-risk, low-complexity, and similar in scope to small tasks. This is where citizen developers commonly add value by automating tasks for themselves or their teams.

3. **Long-running processes** involve multiple digital workers and individuals within the same group or across multiple groups. These processes are designed to be completed over an extended period (hours or days) involving humans in the loop to provide inputs or approvals. These processes' automation generally concentrates on process efficiency and compliance risk reduction.

4. **End-to-end high-value processes** are typically cross-functional processes that span multiple business areas. These are also tied to high-level business objectives addressing cost optimization, revenue growth, and risk reduction.

5. **Complex processes or tasks** usually involve complex business rules and data-driven decision-making using semi-structured or unstructured data. These present opportunities to augment processes by integrating AI-enabled capabilities to unlock strategic value. These are also tied to high-level business objectives addressing cost optimization, revenue growth, and risk reduction.

Establish Preferred Choice of Technology

To lay the robust foundation for a scalable IA ecosystem, IA leaders must build a portfolio of complementary IA technologies based on the organization's priorities. In addition, IA leaders must evaluate various IA products and vendors that align with the organization's business objectives. In this process, IA leaders will undoubtedly encounter competing priorities like cost, time to market, maturity of IA products and vendors, and technical skill level available in the organization.

Furthermore, each business function will present disparate functional requirements and integration limitations. Thus, it's imperative to choose a combination of IA technologies that best fit the needs of business functions. The technology integration facet further covers this portfolio of technologies in detail.

Build a Roadmap for Sustainable Growth

IA leaders must build a long-term strategic roadmap with a vision of wide-scale adoption of IA capabilities across all processes and departments of the organization. The challenges lie in the evolving nature of IA technologies, and thus, IA leaders must determine suitable ways to future-proof the roadmap. Therefore, starting small and incrementally investing in the IA ecosystem is recommended based on the organization's priorities. **Emilie Ly**, Senior Director, BPM & RPA at VMware, Inc., states, "We have found that starting small with several proofs-of-concept to connect business challenges to technology is the best approach to drive intelligent automation adoption. At the same time, it enables us to evolve our strategy and roadmap to scale."

This roadmap must:

1. target to deliver the business value based on the established strategic values.

2. address the iterative steps of discovering process automation opportunities tied to strategic values.

3. cover a holistic view of technology portfolio adoption across the enterprise.

Business Applications

It is important to note that the dependencies of onboarding a business application onto an IA program are still an afterthought for many organizations across all industry verticals.

Each business application presents unique requirements and limitations that must be proactively addressed to prevent critical roadblocks in scaling the IA program. For example, you may be prepared to onboard a business function onto the IA program. However, you may not be able to achieve meaningful outcomes for this business function if its business application store and process personal identifiable information (PII) data and you haven't yet established an approach to handle it with the help of a data privacy team.

Furthermore, business applications are generally not operated stand-alone and thus require integration and data transfer between various other business applications for the business function to perform. Therefore, IA leaders must maintain a holistic view of the usage of business applications and accordingly address the commonly experienced dependencies listed as follows:

1. **Business application owners**: First and foremost, every business application will have an owner either in the IT or business domain. IA leaders must engage with the owners to have an ongoing interlock and active engagement. Most dependencies will become easier once you have established a relationship with application owners.

2. **Application environment setup**: Like software development, IA development requires multiple business application environments to develop and test the bot code. This is a critical requirement to avoid developing and testing the bot code directly in the production environment.

   Moreover, it is not uncommon to encounter that only the production environment is available for specific applications, and IA developers will not have access to lower instances of the application. Therefore, IA leaders must establish an approach by working with IT and business application owners to reduce the risk of having limited environments.

3. **Application access**: The developers and bots should be granted the least possible privilege access to business applications required to develop, test, and run the business process.

   Further, it is recommended to define a process to maintain access for change requests such that any requested changes can be delivered in time without spending unnecessary cycles on granting access again.

4. **Testing**: Testing is a very time-consuming activity while developing IA bots. Moreover, each change requested in the workflow will require the same testing to be performed to ensure that changes have not inadvertently broken a working workflow.

To set up a strong foundation for scalability, IA leaders must consider automating the testing of IA bots. IA leaders must engage with business application owners and SMEs to gather necessary test data to perform automated testing. This testing will allow catching the business application and environment changes in the testing environment before they are released in the production environment.

In addition to the user acceptance test for the bot code, it is recommended to perform performance and load testing to ensure that the frequency of operations performed by bots is adaptable to the business application.

5. **Data security and privacy**: The proper management of the data used by the digital worker is paramount to the success of a scalable IA program. IA leaders must strive to keep the confidentiality, integrity, and availability of data used by the digital worker. It is recommended to establish a process to manage confidential and PII data. This process can start with requirement gathering about process automation by capturing if any confidential and PII data are used in the process or if business SMEs can provide test or mock data for development and testing purposes.

   Further, special techniques must be used for the production environment to handle transient data created during complex manipulation of data exported from the business application. These data must be encrypted, and limited access must be granted to troubleshoot any issue in the production environment.

   Lastly, it is also recommended to establish data security and privacy training modules for developers, business analysts, and support operators to make them aware of the organization's practices and policies.

6. **Regulatory requirements**: These business applications may bring unique regulatory requirements that need to be considered as part of the planning to scale. For example, like knowledge workers, the segregation of duties will be applicable to digital workers, and thus, more than one digital worker will be required to complete certain tasks. Additionally, under European Works Council regulations, organizations may have to obtain works council approval to automate tasks or processes in some countries.

7. **Upgrades and change management**: It's not surprising to state that the majority of bot failures result from a change in business application. A well-defined process must be in place to ensure that business application changes are communicated and tested in time by making necessary changes to the bot code.

8. **Communication**: CoE leadership and support operators must be notified of any activity on the business application which could potentially impact the stability of the digital worker resulting in business process downtime. These activities range from scheduled changes and upgrades, performance and load limitations, and planned maintenance to unexpected downtime.

9. **Business continuity planning (BCP) and disaster recovery (DR)**: IA leaders must ensure that the IA ecosystem is part of the organization's BCP and DR initiative and participate in regular BCP tests. Similarly, any BCP and DR activity triggered on the business application must be notified to CoE to perform any necessary remediation promptly.

Citizen Development

Note that digital transformation has a more important purpose than adding new and emerging digital technologies to business workflows. It facilitates generating value for the organization through innovation; designing faster, compliant, and more accurate processes; and developing more profound insights into customer and competitive data using existing and emerging digital technologies. In addition, by enabling and empowering employees, the citizen development approach can accelerate the adoption of IA capabilities throughout the organization.

The citizen developers are business users who are interested, capable, and available to develop low-complexity and low-risk intelligent automation for themselves and their departments. In addition, the citizen developers are trained to leverage low-code/no-code platforms and can play a critical role in scaling the IA program across the enterprise.

How Can Citizen Development Help Scale?

At the onset, it is essential to note that citizen development is not for everyone. IA leaders must expect to have attrition of citizen developers for various reasons. However, there is still value in training these business users and making them aware of the automation technologies and their capabilities. In addition, these business users can become valuable ambassadors for the IA program in their respective business areas. The goal of citizen development is to enable, equip, and empower business users who best know the business processes, to build intelligent automation they believe is most valuable for their day-to-day activities.

1. **Augment CoE capacity to scale:** Citizen developers can augment CoE capacity by building low-complexity and low-risk automation and have CoE focus on mid-to-high complexity automation.

2. **Low-complexity automation:** Not all automation needs CoE expertise. The citizen developers can build low-complexity automation. This simple automation might become an initial step for the business team to identify more significant opportunities for automation that the CoE can take up.

3. **Employee empowerment:** Empowering business users to partner and build automation can help with the change management process by alleviating the resistance to the adoption.

4. **Productivity gain:** Since business users are closest to the manual activities required to perform a business workflow, they appreciate the value of automating these steps the most. They can identify the most suitable opportunities to automate small tasks to achieve better accuracy and higher employee satisfaction and improve overall productivity.

5. **Identify IA ambassadors**: This program can also help identify employees with the talent and interest in becoming IA ambassadors or specialists.

In the last few years, driven primarily to accelerate the adoption of IA platforms, low-code/no-code capabilities are now available in a range of platforms facilitating the development of application components such as forms, websites, databases, process flows, virtual agents, and AI-enabled capabilities. However, it is important to recognize that low-code/no-code platforms are in their infancy and currently don't have large and experienced communities to help develop best practices. Further, low-code/no-code platforms also involve a learning curve to develop a level of expertise. After all, these platforms include intricate programming constructs such as variables and nested loops.

Governance for Scalability

The decentralized nature of citizen development usually raises concerns about the inability to establish strong governance. For example, IA leaders are concerned about not having enough visibility into what bots are running across the organization, how these bots uphold security and compliance, and how to resolve any issue when the citizen developer is unavailable with the organization.

To scale the IA program enterprise-wide, IA leaders must strive to establish strong governance for the citizen development program.

This governance model must:

1. Address a well-defined scope of automation for citizen development programs. Most organizations prefer to cover low-complexity and low-risk automation that improves employee productivity and satisfaction.

2. Identify the right candidate for citizen development, a tech-savvy business SME interested in learning new skills, able to allocate time for dedicated training and development, and someone with a natural attitude to solve business problems and improve process efficiencies.

3. Provide a well-structured training and enablement program to cover low-code/no-code development capabilities, custom governance practices, and data privacy topics referring to the rules and policies of the organization.

4. Ensure license management that will issue a necessary license to trained developers and provision to revoke the license if it is not used over a period or when unauthorized activities are identified during the audit.

5. Provide business management, SMEs, and CoE oversight to ensure that suitable automation candidates are approved based on complexity and risk.

6. Ensure the selection of the IA platform that supports the control of what activities citizen developers can or cannot perform.

7. Provide CoE oversight to ensure that established best practices are followed for quality code, testing, and production readiness.

8. Offer the ability to monitor and track the usage of the bots for compliance and audit purposes.

Choices for Low-Code and No-Code Development

Many choices are available for low-code/no-code development with significant overlap in capabilities. These tools facilitate enterprise-wide software development with cloud services and can be critical for enterprise-wide innovation. IA leaders must avoid the proliferation of these tools across the organization by selecting a set of low-code/no-code platform combinations suitable for the organization. However, the abundant choices available make it a complex task. Some of them are listed as follows (please note that this is not a comprehensive list):

1. Business process automation platforms to automate from small tasks to long-running processes.

2. Business process management (BPM) platforms for process mapping.

3. Apps development platforms for building Desktop, Web, or Mobile Apps.

4. Website development platforms enable website development with ease, like word processing.

5. Mobile application development for iOS and Android.

6. Artificial intelligence and data science platforms to enable citizen data scientists to build machine-learning models.

7. Data analytics automation platforms for building and automating end-to-end data pipelines and analytics.

8. Blockchain platforms are intended to allow using blockchain technology in the application without going deeper into the nuances of blockchain.

9. IoT application development platforms that enable building platform-agnostic applications.

Technology Integration

The central concept of true IA is in combining and integrating existing and emerging IA technologies. These integrations are essential to bringing efficiency to business processes and improving data-driven human decision-making.

IA technologies are rapidly evolving. Therefore, the IA implementers and end-users must detach from the notion that bots can only perform simple tasks like copying and pasting data, opening and moving files, and extracting and storing data. Instead, IA capabilities must be looked at from a broader perspective. This is where the "intelligence" of IA plays a vital role in augmenting business processes with AI-enabled capabilities. This is where the full potential of IA technologies can make a significant difference in making existing business processes efficient, accurate, and sustainable by integrating technologies like RPA, ML, NLP, IDP, chatbots, and many more.

You may discover use cases aligned with the organization's priorities whereby usage of IDP can result in significant strategic value, you may encounter scenarios whereby introducing chatbots can improve process efficiencies resulting in customer satisfaction, and your organization might already have a matured data science team who has built complex ML models to address critical business use cases which can be great candidates to augment and automate business process. The point is that there is no one-size-fits-all approach to adopting IA technologies. It will vary significantly from organization to organization and the journey they choose to take to achieve specific business objectives and goals (Beyond RPA: Build Your Technology Portfolio for Hyperautomation, n.d.).

In the context of the scalability of the IA program, integrating IA technologies will be an iterative and evolving process. Therefore, a well-defined roadmap is needed to roll out the building blocks of the IA technologies for enterprise-wide adoption. This adoption strategy must outline a path forward based on the organization's objectives, priorities, and funding.

Building Blocks of IA Technologies

This section presents an overview of existing and emerging IA technologies with a motivation of introducing the technology, its application, and factors to consider while integrating. It is essential to consider the evolution and integration of these technologies agnostic to IA platforms that implement them. You may encounter that a given IA platform implements or supports one or more of these technological capabilities. At the same time, one platform might be dedicated to a specific technology. This section covers the list of technologies (see Figure 7.2) to provide a flavor of the IA technology landscape. The following section, about architecture and infrastructure facet, covers the relevant platforms constituting the IA ecosystem. Also, note that IA technology is an ever-evolving field; thus, the list below may not be exhaustive.

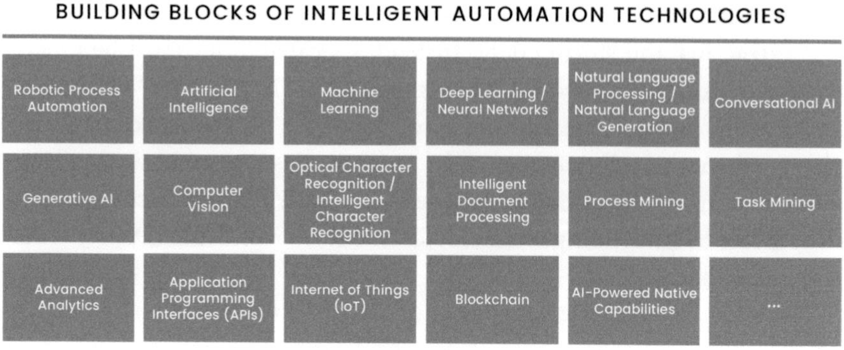

BUILDING BLOCKS OF INTELLIGENT AUTOMATION TECHNOLOGIES

Robotic Process Automation	Artificial Intelligence	Machine Learning	Deep Learning / Neural Networks	Natural Language Processing / Natural Language Generation	Conversational AI
Generative AI	Computer Vision	Optical Character Recognition / Intelligent Character Recognition	Intelligent Document Processing	Process Mining	Task Mining
Advanced Analytics	Application Programming Interfaces (APIs)	Internet of Things (IoT)	Blockchain	AI-Powered Native Capabilities	...

FIGURE 7.2   Building Block of Intelligent Automation Technologies.

Robotic Process Automation (RPA)   RPA helps to build software robots, also known as bots, that emulate human actions of typing using the computer keyboard, navigating software systems screens, recognizing and extracting the data on the screen, etc. These bots can be leveraged to automate rule-based mundane and repetitive tasks that can be time-consuming, labor-intensive, and error-prone. The RPA bots are ideal for automating legacy systems processes.

One of the key benefits of RPA is that compared to other automation technologies, these bots can be implemented faster and easier to handle business process changes, making them more adaptable than traditional automation solutions. In addition, these bots can perform all actions on a screen like humans but faster, more accurately, and without a break. Additionally, these bots can be augmented with AI capabilities making them efficient digital workers.

Artificial Intelligence (AI)   AI is the decision engine for IA. AI is the foundation for mimicking human intelligence by applying software algorithms. Essentially, AI makes computers think and act like humans. AI is further classified into three main categories.

Artificial narrow intelligence (ANI), also known as weak AI, is skilled at one specific task but tends to outperform humans in the task it has been trained on. At present, the most common usage of AI falls under this group. For example, recommendation engines on Amazon or Netflix, image recognition for medical diagnosis, predictive algorithms for the stock market, digital assistants like Alexa or Siri, and even self-driving cars like Tesla are all applications of the ANI.

Artificial general intelligence (AGI), also known as strong AI, is meant to enable a machine's capability to perform a wide array of human-like tasks. This doesn't exist at present. Although the excitement around the ultra-powerful chatbot ChatGPT felt like the advent of AGI, it is far from an AGI-level system.

Artificial super intelligence (ASI), also known as hypothetical or imaginary AI, refers to when the capability of computers surpasses human intelligence and behavioral ability, which is currently science fiction.

Machine Learning (ML) ML is a subfield of AI, broadly defined as the capability of machines to learn and imitate intelligent human behavior without explicitly programming (Brown, 2021). There are mainly three types of ML.

Supervised learning needs external supervision to train the ML models. This supervision is provided using labeled data to train ML models. The supervised learning algorithms then map labeled inputs to the known outputs. This is the most commonly used ML method applied for cases like stock price analysis, weather prediction, and sales forecasting.

Unsupervised learning uses unlabeled data to train machine learning models. These models learn from the data to discover the data's patterns and trends (features). No external supervision is required in this training process. These models learn on their own and return the output. One of the applications is customer segmentation, whereby similar customers can be grouped into clusters based on their likes, dislikes, and behavior.

Reinforcement learning trains a machine to take suitable actions and maximize its rewards in a particular situation. It follows trial-and-error methods to get the desired result. This method also does not need any external supervision to train models. These algorithms are widely used in the gaming industry to build games and also to train robots to do human tasks.

Deep Learning/Neural Networks Deep learning is a subfield of ML inspired by the structure and function of the human brain. Deep learning algorithms use artificial neural networks to identify patterns within unstructured data (Brownlee, 2020). Like the human brain, these artificial neural networks are organized in layers consisting of interconnected neurons. Each neuron is a function that transforms input data to produce the result through the final (output) layer.

Deep learning has gained popularity in the past decade as current computers are powerful enough to use these techniques properly. Deep learning is ideal for complex tasks where a large set of unstructured data is required for training. Some well-known deep learning applications include image recognition, speech recognition, fraud detection, and medical diagnosis.

Natural Language Processing (NLP)/Natural Language Generation (NLG)
NLP is a field of AI that allows machines to read, understand, and derive meaning from human languages. For example, NLP enables machines to understand the text from emails, chatbot interactions, social media posts, etc. Similarly, NLG generates text and speech from data (Yse, 2019).

For IA leaders, there are abundant opportunities to integrate NLP/NLG capabilities into business processes. Sentiment analysis can be performed on social media textual data to help organizations better monitor customer feedback and understand their needs. Machine translation is another powerful use case for organizations with a global presence. This technique allows organizations to translate text from a source language to a target language and share documentation, like training modules, with international customers, partners, and employees.

Conversational AI   Conversational AI refers to AI technologies that enable chat messaging and speech-enabled human-like interactions with computer software applications. Conversational AI is used for user-centric processes where user engagement involves messages (text, email, web, etc.) or speech (phone, etc.). It uses NLP/NLG algorithms for sentiment analysis and speech analytics to better understand the conversation.

There are many opportunities where automating simple FAQs, knowledge queries, and service requests can be managed with the help of chatbots and virtual agents where employees and customers will have human-like interactions with these bots/agents freeing up valuable service desk staff to manage requests which indeed requires human interactions (Sharma, 2022). Further, integrating chatbots with RPA robots creates opportunities to perform tasks/processes through text or speech communication channels.

Traditionally, conversational AI is adopted for customer service to improve the customer experience for use cases like bank loan applications or insurance claims. It also offers similar opportunities for IT, HR, or Legal services, to improve the employee experience.

Generative AI   Generative AI refers to the category of AI algorithms or models designed to generate new content such as text, images, audio, videos, programming code, and synthetic data. Their use generally starts with a prompt containing text, image, audio, video, or any type of input that the model can process. The Generative AI model then returns new content in response to this prompt. This new content can be a description of a solution to a problem, a drawing of product design, new musical notes, a video for a marketing campaign, or a realistic-looking fake image.

Currently, there are two widely used Generative AI models.

a. **GANs (Generative Adversarial Networks):** The GANs consist of two different types of neural network models – a generator and a discriminator. The generator generates new data, whereas the discriminator is trained to differentiate the real data from the data generated by the generator. During the training of these models, the generator repeatedly tries to produce data to "trick" the discriminator and adapt to the results that are successful. This training process continues until the generator wins this "adversarial" contest by producing data that are entirely indistinguishable from the real data. This "adversarial" process aids in the improvement of both generator and discriminator models, eventually resulting in higher-quality generated data.

b. **Transformer-based models**: The transformer-based deep learning model, such as BERT (Bidirectional Encoder Representations from Transformers) and GPT (Generative Pretrained Transformer), learns by capturing the relationships between different elements of sequential data. For example, these models can derive the meaning and context from long sequences of text by recognizing the relationships (order, proximity, etc.) between different words or semantic components of a language. This allows these models to understand natural language better and generate realistic, high-quality text.

Generative AI is not a new technology. The GANs have been around for approximately a decade. However, Generative AI became part of the mainstream conversation due to the recent advancements of large language models (LLMs) and the release of ChatGPT by OpenAI in November 2022. With the new advancements, Generative AI has the potential to disrupt many industries with a wide range of practical applications across all industries, such as scenario generation for risk assessment and management, generating options for business process optimization, rapid creation of assets for marketing campaigns, generating simulations for new drug discoveries, generating product designs and engineering drawings, and many more.

Computer Vision   Computer vision is a collection of deep learning techniques, such as object detection and image classification, enabling machines to interpret and understand visual information from images, documents, screens, and videos. RPA and IDP platforms are fundamentally developed using computer vision techniques.

Optical Character Recognition (OCR)/Intelligent Character Recognition (ICR)
Optical character recognition (OCR) is a process to recognize and convert the text in an image file, such as photos or scanned documents, into a machine-readable text format. Similarly, intelligent character recognition (ICR), also known as intelligent OCR, is used for recognizing and converting handwritten text, or fonts and different styles of handwriting, from image files.

Intelligent Document Processing (IDP)   The IDP platform provides capabilities to convert unstructured and semi-structured data from documents into structured and usable information. These capabilities, integrated with RPA bots, can be beneficial for automating document-centric business processes.

The IDP typically makes use of many technologies, including AI, ML, deep learning, NLP, computer vision, OCR, and ICR, to read and understand business documents like invoices, purchase orders, tax documents, legal contracts, and insurance claims.

Process Mining   Process mining is a technique designed to discover, monitor, and improve real processes (i.e., not assumed processes) by extracting readily available knowledge from the event logs of information systems (Definition of Process Mining – Gartner Information Technology Glossary, n.d.).

Process mining is a very effective enabler for IA. It extracts relevant information, such as process execution data (date, timestamp, user, activity, etc.), from event logs of the information system to visually represent the process. It enables organizations to gain a deep insight into their business processes, involving process exceptions and anomalies, and can be critical in simplifying and optimizing them. It helps identify process areas, such as repetitive, time-consuming, or error-prone tasks, that can be automated. In addition, it can also be leveraged to monitor and measure the effectiveness of automation by providing a comprehensive view of how processes are executed before and after the implementation of automation.

Generally, there are three levels of process mining.

1. **Discovery**: This method uses event logs to produce a process map without any assumptions or external information about the process.

2. **Conformance**: This method confirms whether the process map conforms to the process followed in practice. It helps in identifying any outliers or deviations from the intended process.

3. **Enhancement**: This method enhances or improves the discovery and conformance model by applying an improvement from a proposed process map.

Task Mining   Task mining is a technique designed to collect and explore user data (user's clicks, keystrokes, and other interactions) about real-time interactions with the front end of the business application. Typically, this is done by using computer vision techniques.

These user data are used to generate process maps that show exactly how a process is being executed by the user, highlighting various decision points and variations. Anonymization techniques are used to mitigate privacy concerns regarding PII and other sensitive data. Further, ML algorithms are used to generate insights about the user process's efficiency, optimization, and automation.

It is important to note that using task mining and process mining together provides a complete view of how processes are truly performed.

Advanced Analytics   Advanced analytics is the data analysis method that uses statistical methods, ML and deep learning algorithms, and business process automation beyond the capabilities of traditional business intelligence (BI) tools to analyze data to gain deeper insights into the business.

"Advanced Analytics is the autonomous or semi-autonomous examination of data or content using sophisticated techniques and tools, typically beyond those of traditional business intelligence (BI), to discover deeper insights, make predictions, or generate recommendations. Advanced analytic techniques include those such as data/text mining, machine learning, pattern matching, forecasting, visualization, semantic analysis, sentiment analysis, network and cluster analysis, multivariate statistics, graph analysis, simulation, complex event processing, neural networks."

Definition of Advanced Analytics - Gartner
Information Technology Glossary (n.d.)

Advanced analytics enables the true potential of IA through the analysis of structured, semi-structured, and unstructured data. The unstructured data lack any structure and have no associated data model, making it very difficult to analyze and manage. The texts (emails, chats, social media posts), images, audio, videos, and IoT sensor data are unstructured data. Semi-structured data are partially structured that refer to the same type as unstructured data, but they also contain metadata that define the characteristics and partial structure of the data. In contrast, structured data consist of well-defined data types.

IA not only automates but also digitizes business processes, which in turn generates more data. Using advanced analytics, these data can help better understand business processes and workflows, help drive process optimization, and potentially gain unprecedented insights into business processes.

APIs (Application Programming Interfaces) The user interface (UI)-driven IA can sometimes be fragile, especially for applications prone to UI changes. Some IA platforms support native API integration, allowing integration directly with the back end of enterprise applications. These APIs enable software systems to communicate with each other using a set of predefined interfaces and protocols. As a result, APIs are a very effective way to access data from disparate systems. In addition, most cloud providers, such as AWS, Azure, and GCP, provide cloud-based AI APIs. These APIs can also be integrated with business workflows to develop IA.

Internet of Things (IoT)   "The Internet of Things (IoT) is the network of physical objects that contain embedded technology to communicate and sense or interact with their internal states or the external environment" (Definition of Internet of Things (IoT) – Gartner Information Technology Glossary, n.d.).

In the past decade, an overflow of structured, semi-structured, and unstructured data was generated due to the proliferation of IoT devices. The consumption and processing of sheer volume, velocity, and variety of these data are beyond humans' capacity, and this is where strategic integration of IoT with other IA technologies like AI/ML, RPA, NLP, and Blockchain can create opportunities to exploit and use these novel data sources. Such integration could effectively automate monitoring and managing the IoT ecosystem, enabling collecting, cleansing, and analyzing data to uncover unprecedented insights, detect precise anomalies, and even propose appropriate remediation.

For example, integrating IoT, AI/ML, and advanced analytics can dramatically improve manufacturing quality. One commonly used scenario is where IoT sensors monitor the manufacturing process, AI/ML algorithms detect defects in the production, and advanced analytics is used to improve and optimize the manufacturing process to reduce defects. Similarly, integrating IoT with other IA technologies creates opportunities for improving preventative maintenance of critical healthcare, manufacturing, military, or smart grid equipment.

Blockchain   Blockchain is a decentralized ledger that records immutable transactions securely and transparently. The transaction is saved on a "block" linked to another block forming a chain called Blockchain. Formally:

> A blockchain is an expanding list of cryptographically signed, irrevocable transactional records shared by all participants in a network. Each record contains a time stamp and reference links to previous transactions. With this information, anyone with access rights can trace back a transactional event, at any point in its history, belonging to any participant. A blockchain is one architectural design of the broader concept of distributed ledgers.
>
> *Definition of Blockchain - Gartner Information*
> *Technology Glossary (n.d.)*

Integrating Blockchain with other IA technologies, like RPA and AI, has significant potential and is a natural path of technological evolution that is currently not fully explored. Very limited organizations have successfully attempted this integration. At the same time, many organizations are in the initial stages of their integration journey.

For example, integrating RPA, AI, and Blockchain, where RPA bots automate business data gathering and blockchain stores the data that are used to support AI, can be a very common scenario of intelligent automation.

1. RPA bots can gather structured, semi-structured, and unstructured data for the given business use case. These bots can use screen scraping (computer vision), IDP, and API calls to gather these data from multiple sources.

2. Further, blockchain stores data over a trusted decentralized network of computers. The inherent immutability and transparency of data stored in blockchain improve the authenticity of the data.

3. AI/ML models use this authentic and trusted data to assist in complex decision-making. Further, this trustworthy and transparent data can be used to improve AI-enabled systems.

Such integrations create opportunities to develop scalable solutions for automated and trusted data processing needed for regulatory and compliance management, supply chain management, payments processing, fraud prevention, insurance claims processing, and many more.

AI-Powered Native Capabilities   Many enterprise application software platforms, like Customer Relationship Management (CRM), Enterprise Resource Planning (ERP), Human Capital Management (HCM), IT Service Management (ITSM), provide AI-powered native capabilities. Typically, these platforms support AI-enabled workflow automation, scheduling, analytics, and process orchestration capabilities. IA leaders must consider these native capabilities to incorporate into respective business processes as, in many cases, these capabilities will be more effective and scalable than integrating with non-native capabilities.

Architecture and Infrastructure

The IA architecture and infrastructure refers to the IA ecosystem that consists of several tools, platforms, and infrastructure components. The IA architecture and infrastructure should allow deploying IA enterprise-wide to establish a scalable IA program that can support steady growth and rapid changes. For a scalable IA program, the IA architecture and infrastructure should be able to support the increasing number of bots, the growing number of transactions processed by these bots, and the rising number of hardware nodes to deploy and run these bots.

With reference to the Ecosystem Architecture Concept described in the previous chapter (see Figure 6.12), the architecture and infrastructure discussed here primarily address the scalability of the Intelligent Automation Core Platform and Adjacent Technology Components. This section will also address the dependency on other platforms (like IAM). Note that it is not addressing the components of Enterprise Applications and Enterprise Collaboration. Additionally, treat the ecosystem depicted in Figure 7.3 as a reference architecture. It might evolve differently in your organization depending on the sequence of adoption and integration of the building blocks of IA technologies discussed in the previous section.

To build a scalable IA architecture and infrastructure, **Julio Viquez**, Senior Manager, Intelligent Automation at VMware, Inc., cautions, "be mindful that with the growth of the IA program, the complexity of the infrastructure and architecture is going to grow as well. Thus, it is prudent to invest in simplifying and automating infrastructure administrative

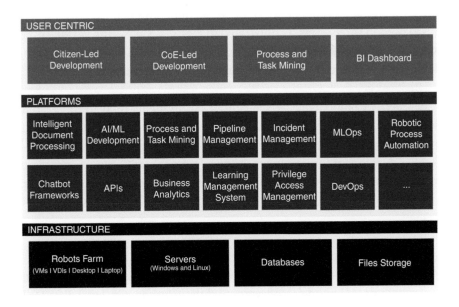

FIGURE 7.3   Intelligent Automation Technology Ecosystem.

tasks early on." Julio further advises keeping an inventory of the technical debt incurred over time to address them in a timely manner to avoid roadblocks to scale.

Thus, it is essential first to recognize:

1. What are the software and hardware components required for the IA ecosystem?

2. What kind of deployment environments needs to be set up and managed?

3. What deployment models (on-premises, cloud, or hybrid) are available to set up?

4. What skills and personas are needed to manage a scalable infrastructure?

The previous section talked about a list of IA technologies, their capabilities, and how integrating them provides the true value of IA. This section covers the IA ecosystem comprised of various platforms that implement these IA technologies and assist in managing the IA program.

Components of Intelligent Automation Technology Ecosystem

The IA technology ecosystem can be understood and managed well when it is viewed in three levels, as illustrated in Figure 7.3, covering various user tools, software platforms supporting capabilities needed to implement, deploy, and manage IA, and infrastructure to deploy these platforms.

User-Centric Components User-centric components refer to integrated development environment (IDE), analytics tools and dashboards used by IA developers, business users, and SMEs to analyze business processes, develop automation workflows, prepare and report strategic matrices about the benefits and growth of the IA program, and monitor and maintain operational matrices concerning the IA ecosystem and its usage.

1. **CoE development tools:** These tools are required for software development meant for the implementation of RPA and IA workflows, chatbots, ML models, software modules (Python, C#, Java), RESTful APIs, scripts (Python, PowerShell), and many more.

2. **Citizen development tools***:* These low-code/no-code tools enable and empower business users to automate repetitive tasks and develop apps without writing code by using intuitive point-and-click user interfaces for easy and faster development. In addition, these tools offer integration with various business applications and pre-built AI-powered capabilities.

3. **Process mining and task mining:** These user-centric components of process mining and task mining allow business users and process engineers to perform process analysis and discovery. The resulting insights presented through visualizations and analytics are used for process standardization, optimization, and automation.

4. **Business intelligent (BI) dashboard***:* The dashboard encompasses data visualization and analysis tools that can be used for various purposes. These dashboards can help capture and present KPIs aligned with business objectives. CoE can also develop dashboards for monitoring and reporting project status and pipeline, incident management, bot utilization, infrastructure footprint, growth, stability, and many more.

Software Platforms Components   The software platform layer comprises a collection of wide-ranging products and platforms (see Figure 7.4) needed to scale the IA program enterprise-wide. This layer broadly comprises software products that implement IA technologies discussed in the previous section and software products needed to run the IA program effectively.

Technology Platforms   Here we will discuss various platforms that implement IA technologies. The key challenge for IA architect is identifying a suitable software product for the organization. In addition, IA architects must ensure that the correct configuration is set up for each platform to provide a scalable, highly available, and secured environment. Many solutions will span multiple platforms. The IA architects should develop a holistic view of these platforms' usage, capacity, capability, and limitations.

1. **Robotic process automation (RPA) platform***:* The RPA platform provides capabilities that make it easier to develop, deploy, and manage software robots. In addition to providing a user interface to

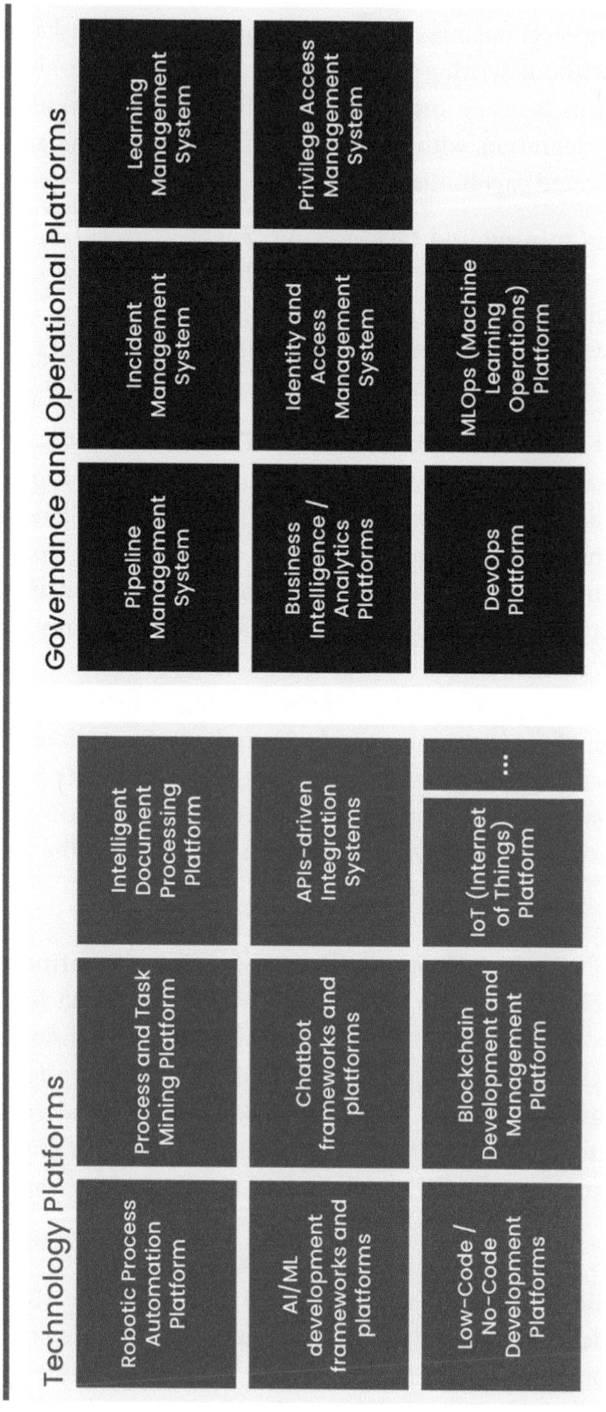

INTELLIGENT AUTOMATION ECOSYSTEM

Technology Platforms

Robotic Process Automation Platform	Process and Task Mining Platform	Intelligent Document Processing Platform
AI/ML development frameworks and platforms	Chatbot frameworks and platforms	APIs-driven Integration Systems
Low-Code / No-Code Development Platforms	Blockchain Development and Management Platform	IoT (Internet of Things) Platform
		...

Governance and Operational Platforms

Pipeline Management System	Incident Management System	Learning Management System
Business Intelligence / Analytics Platforms	Identity and Access Management System	Privilege Access Management System
DevOps Platform	MLOps (Machine Learning Operations) Platform	

FIGURE 7.4 Intelligent Automation Technology Ecosystem.

develop the bots, these platforms also provide ways to communicate with business systems through screen scraping or API integration. Additionally, these platforms also offer control panels to monitor and manage the robots and operational dashboards presenting insights into resource usage and transaction processing.

2. **Intelligent document processing (IDP) platform**: The IDP platform provides capabilities to convert unstructured and semi-structured data from documents into structured and usable information. These capabilities, integrated with RPA bots, can be beneficial for automating document-centric business processes. These platforms typically combine OCR and NLP capabilities to read and understand business documents like invoices, purchase orders, tax documents, legal contracts, KYC documents, insurance claims, and many more.

3. **Chatbot frameworks and platforms**: A chatbot uses conversational AI technology to simulate and process human conversation via text or voice. A human can interact with chatbots through multiple channels like messaging applications, mobile apps, webpages, or phones. Many choices are available for these frameworks and platforms, offering support to develop chatbots for wide-ranging communication channels (web, mobile, etc.) to specialized channels (voice only, social media integration only, etc.). Other offer advanced capabilities like support for a large set of human languages, deep dialogue context analysis, or video interaction.

4. **Application programming interfaces (APIs)**: Centralized API platforms can enable organizations to access data from anywhere, whether data reside on-premises or in the cloud. In addition, these platforms can facilitate data access policies to manage, monitor, and secure data at scale for IA. Further, most cloud providers, such as AWS, Azure, and GCP, provide cloud-based AI APIs. These APIs can be integrated with business workflows to develop IA such as image and video analysis, fraud prevention, demand forecasting, and many more.

5. **AI/ML development frameworks and platforms**: These frameworks and platforms enable data scientists, data engineers, and ML engineers to build ML projects from ideation to production by offering building blocks to design, train, validate, and deploy ML models. Many choices are available for these frameworks and

platforms, offering support to develop wide-ranging models to specialized (conversation AI, etc.) models. Some of these frameworks are open source offering flexible architecture to deploy models on CPUs or GPUs on desktops, laptops, servers, or mobile devices. Other offers cloud-based advanced analytics and APIs to simplify ML development. In addition, some enterprise AI platforms enable democratizing AI/ML model development, empowering citizen data scientists to build low-complexity models.

6. **Process and task mining platform**: These platforms assist in uncovering the entire business process by examining event data from transaction systems and user activity data through an agent deployed on the user's workstation. These platforms enable process discovery and mapping to represent process steps visually. It also allows a thorough analysis of the end-to-end process to help identify process improvement and optimization areas. Further, it facilitates ongoing monitoring of the processes and alerts when a deviation is identified. In addition, these platforms' process enhancement and simulation capabilities allow for exploring the impact of proposed changes before changing the entire business process.

7. **Blockchain**: These platforms provide frameworks to develop blockchain-powered applications. These frameworks enable the development of public, permissioned, and private network blockchain applications. Public networks are decentralized networks that any participant can join. Permissioned networks are visible to the public, but their participation is controlled. Private networks are not open to the public and are only used by trusted parties.

8. **IoT**: The IoT platforms, including software and hardware, allow the development of new applications, offer IoT devices and connectivity management, support secure data storage, and assist in advanced analytics. These platforms are classified into four categories.

   a. **Application enablement**: The application enablement platforms enable the development of IoT applications and solutions such as smart home devices or industrial control systems.

   b. **Device management**: Device management platforms provide software and hardware building blocks that monitor, troubleshoot, and update the IoT devices remotely.

c. **Connectivity**: The connectivity platforms allow monitoring and managing the connectivity of the IoT technology stack across communication protocols like Bluetooth, Wi-Fi, and cellular technologies like 3G/4G/5G, or Narrowband IoT (NB-IoT).

d. **Analytics**: The analytics platforms enable using the data generated by IoT devices to develop actionable insights.

9. **Low-code/no-code development platforms:** These platforms reduce software development complexity by leveraging visual building blocks such as drag and drop. In addition, these platforms use techniques like code templates, plug-ins, graphical connectors, and APIs to automate a significant portion of the development process. As a result, these platforms effectively enable citizen developers to build desktop/mobile/web apps or automate business processes without a formal programming background. In addition, some enterprise AI platforms enable democratizing AI/ML model development, empowering citizen data scientists to build low-complexity models.

Governance and Operational Platforms

1. **Pipeline management**: To achieve enterprise-wide IA, you must have a way to manage and track the entire life cycle of every automation. This platform enables centrally managing and prioritizing automation ideas and opportunities. In addition, the centralized process repository enables effective reporting of business matrices and KPIs.

2. **Incident management**: This platform captures any failures or abnormalities of bots resulting in downtime and assists in analyzing, resolving, reporting, and auditing these incidents. This platform is essential for building a strong foundation to scale IA practice by capturing common failures, common changes, SLA breaches, incident priorities, etc. These data points can be strategically used to improve the bot's quality and reduce failure rates.

3. **Business intelligence/analytics**: This tool offers data analysis and visualization capabilities to build performance scorecards capturing and reporting business metrics and KPIs. Many IA core platforms offer integrated solutions to develop these dashboards; however,

some organizations might prefer to use a platform strategically aligned with enterprise-wide business analytics.

4. **Learning management system (LMS)**: On-going awareness of growing IA capabilities and upskilling the CoE, business users, and other involved groups is critical for the successful scalability of IA practice. The LMS tools allow the life cycle management of learning and development programs by enabling training programs' administration, documentation, delivery, tracking, and reporting.

5. **Access management systems**: Identity and access management (IAM) and privilege access management (PAM) are two access management systems used to manage identity authentication and authorization. The bots need privileged access to critical business applications with sensitive data to run critical business processes. These bots should be assigned their own bot accounts – a service account or an equivalent bot account – and granted credentials (Bot ID, passwords, and/or keys) with the least privileged access to these applications. The IAM and PAM tools allow the life cycle management of the bot's IDs, role-based access levels, and credentials management.

6. **DevOps**: The DevOps tools and practices assist in developing and orchestrating end-to-end bot life cycle management with continuous testing and automated deployment. These tools are targeted to integrate and automate IA development and operations activities, resulting in improved robot delivery, quality, and scalability. These practices enable the collaboration between automation engineers, automation operations analysts, and business users. For scaling the IA program, it is essential that this process is lightweight, flexible, and customizable.

7. **MLOps (machine learning operations)**: These platforms offer frameworks and environments that support ML life cycle management. These platforms enable collaboration between Data Science, IT, and CoE teams to accelerate the pace of ML model development and deployment. In addition, these tools help streamline the process of deploying ML models to production and then maintaining and monitoring them, which is essential for scaling the IA program.

Hardware Infrastructure Components   The infrastructure layer encompasses hardware components required to deploy and manage components of IA ecosystems like software platforms, databases to support platforms, ML models, and a pool of bots. This layer is a collection of multiple infrastructure components like virtual machines (VMs), virtual desktop infrastructure (VDIs), laptops and desktops, Windows and Linux servers, Database servers, and shared file storage. Depending on the level of integration required, the physical architecture to support the IA ecosystem could rapidly become very complex, involving load-balancers, in-memory data storage, containerized deployments, etc.

1. Desktop/laptops for bot development, testing, and deployment: Primarily Windows-based devices are used for this purpose, but some vendors have started supporting cross-platform deployments to support other operating systems, like macOS.

2. Virtual machines (VMs)/virtual desktop infrastructure (VDIs) for bot development, testing, and deployment: Primarily Windows-based devices are used for this purpose, but some vendors have started supporting cross-platform deployments to support other operating systems, like macOS.

3. Windows and Linux VMs or bare-metal servers for deploying key server-side components of the IA ecosystem: In addition to deploying technology and governance/operational platforms discussed in the previous section, these servers will be required to deploy and manage components like load-balancers and in-memory data storage for scalable platform deployment and a containerized cluster (Docker, Kubernetes) nodes for scalable deployments of bots and platforms.

4. Series of databases for running and managing IA platforms: Most of the platforms require a database in the backend. While setting up these databases, the IA architects must consider best practices covering data isolation, security, redundancy, backup, access control, etc.

5. Multi-node deployments of IA ecosystem software platforms for high availability: These deployments will require additional hardware, like servers, databases, and file storage, to deploy and manage redundant nodes to achieve high platform availability. Typically, such investment is made for the production environment. However,

multi-node deployment can also be considered for the stage environment to ensure load and performance testing is conducted in a production-like environment.

6. Specialized hardware (CPUs/GPUs) might sometimes be required to build and train ML and deep learning models. GPU acceleration is a key driver and dominating performance influencer for rapid ML development. Before developing a ML model, a significant effort is typically applied for data analysis and cleansing on multi-core CPUs. Thus, the optimal hardware configuration will depend on your specific requirement of the ML project.

7. Similarly, high-performance computers (HPCs) might be required depending on the blockchain project's requirements.

Deployment Environments and Models   It is strongly recommended to have multiple environments of the IA ecosystem. These environments should be set up similarly to the classic software development environments – development, test/stage, and production. It is also recommended to have an optional sandbox environment that allows developers, architects, and business users to experiment with and learn new features and tools/platforms of the rapidly growing IA ecosystem. Alternatively, the development environment can be used for this purpose. The independent sandbox environment prevents any disruption to development and testing performed in development and test/stage environments. Please refer to Figure 7.5 for details about the usage of these environments. All these environments will be set up with a distinct set of VMs/VDIs, servers, databases, and appropriate role-based access. The choice of deployment on-premises or cloud can be a critical decision for the scalability of the infrastructure.

Deployment Models   Most of the platforms in the IA ecosystem support multiple deployment models offering the choice to opt for the model that best suits your organization.

1. **On-premises/private cloud**: This option deploys all components in the organization's data center or private cloud managed by the IT. This option grants maximum flexibility, control, customization, and

DEPLOYMENT ENVIRONMENT OPTIONS

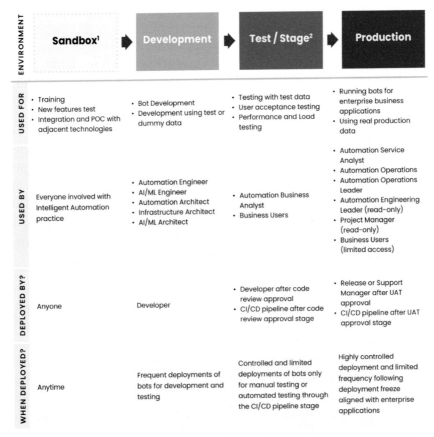

ENVIRONMENT	Sandbox[1]	Development	Test / Stage[2]	Production
USED FOR	• Training • New features test • Integration and POC with adjacent technologies	• Bot Development • Development using test or dummy data	• Testing with test data • User acceptance testing • Performance and Load testing	• Running bots for enterprise business applications • Using real production data
USED BY	Everyone involved with Intelligent Automation practice	• Automation Engineer • AI/ML Engineer • Automation Architect • Infrastructure Architect • AI/ML Architect	• Automation Business Analyst • Business Users	• Automation Service Analyst • Automation Operations • Automation Operations Leader • Automation Engineering Leader (read-only) • Project Manager (read-only) • Business Users (limited access)
DEPLOYED BY?	Anyone	Developer	• Developer after code review approval • CI/CD pipeline after code review approval stage	• Release or Support Manager after UAT approval • CI/CD pipeline after UAT approval stage
WHEN DEPLOYED?	Anytime	Frequent deployments of bots for development and testing	Controlled and limited deployments of bots only for manual testing or automated testing through the CI/CD pipeline stage	Highly controlled deployment and limited frequency following deployment freeze aligned with enterprise applications

1. Setup of Sandbox environment is optional. These activities can potentially be performed on Development or vendor provided environment

2. Test and Stage environment can be setup as two separate environments whereby staging is primarily used for UAT (user acceptance testing)

FIGURE 7.5 Deployment Environment Options.

reliability. It also provides high security and data privacy, as only authorized persons from IT and CoE will have access. However, all these benefits come with high CapEx (Capital Expenses) costs. In this option, CoE and IT must manage the deployments, which could be a significant effort and cost.

2. **Public cloud**: In this option, the components of the IA ecosystem are deployed on the equipment (VMs, servers) provided and managed by public cloud service providers (AWS, Azure, GCP, etc.). This option

offers reduced infrastructure management as most of it is performed by the service provider, reduced costs (no CapEx – Capital Expenses, only OpEx – Operational Expenses) due to the pay-as-you-go model, and high scalability as additional capacity can be easily added. However, these benefits come with risks to data security and privacy. In this case, CoE and IT will support administrative activities with minimal effort for infrastructure management, upgrades, and security patches.

3. **Software-as-a-Service (SaaS) deployment:** Many vendors of IA platforms have started offering SaaS deployments managed by vendors on their cloud. CoE and IT are removed from day-to-day infrastructure management, periodic upgrades, and security patches in this case. Vendors manage these activities. CoE and IT provide an administrative role to grant/revoke access, alert monitoring, license management, etc. However, these benefits come with reduced reliability and risks to data security and privacy.

4. **Hybrid:** This deployment allows the flexibility of having a combination of the above three options. In the context of the IA ecosystem, where we deal with critical business flows and sensitive customer and employee data, you would have a hybrid deployment where some components are deployed on-premises for security and others on the cloud for flexibility and scalability.

Establish Automated Pipeline   Continuous integration and continuous deployment (CI/CD) pipelines must be established. These pipelines can be triggered either on a predefined schedule (daily, weekly, etc.), when a code change is committed, or both. It is advisable to consider both options to ensure periodic testing of the automation, especially when it doesn't change for a more extended period (which is typical for IA if the underlying process is stable). The periodic run of the pipeline ensures that errors are caught early in the lower environment, giving us time to react before underlying application change is made in the production environment.

It is strongly recommended to keep this process lightweight and flexible. Figure 7.6 depicts an example pipeline with steps covering code commit to the code repository, code reviews consisting of both automated and manual checks, automated testing with steps to review the results, and automated approvals triggers for deployment to test/stage and production environments. Table 7.1 lists various pipeline activities with the roles who

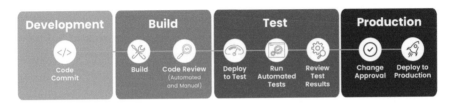

FIGURE 7.6 Steps of the Automated Pipeline.

TABLE 7.1 Roles and Activities for the Automated Pipeline

Activity	Developer	Lead	Operations	Business SME
Code commit	Y			
Build	Y			
Automated code review	Y	Y		
Deploy to test	Y	Y		
Run automated tests	Y	Y		Y
Production release approval		Y	Y	Y
Deploy to production			Y	

perform these activities. Prior to establishing and activating the pipeline, the IA architects must set up the following dependencies:

1. Set up code repository/version control system. You should have your project source code and dependent artifacts committed in a source code repository. This will first require identifying the choice of code repository (Git, Gitlab, etc.) used in your organization.

2. Setup of DevOps environment in a platform of your choice (Jenkins, Azure DevOps, etc.). Work with IT to either leverage one of the existing environments or set up an isolated DevOps environment for IA projects.

3. Establish coding best practices that drive standardization across all developers, such as naming convention and logging standards, removing unused code, using reusable components, and preventing hardcoding credentials, logging credentials or PII data, etc.

4. Establish an automated code review process that generates every code review report highlighting code coverage percentage, violations, and warnings based on pre-established coding best practices. The primary

goal of automated code review is to minimize manual code review efforts, not eliminate them. Automated code review should not pass in case of any violation. All warnings should be resolved as well. However, the warning can be manually reviewed in case of exceptions.

5. Establish a framework for automated testing to test bot code in a repeatable manner. This might require refactoring existing code to perform automated testing effectively. Engage with business SMEs to establish a process for preparing test cases and test data.

6. Establish an approval process for production releases using a check-list, including code review and testing results, with consideration of manual approvals to handle exceptions.

Key Considerations IA architects and infrastructure engineers are encouraged to consider the following recommendations for effectively managing these deployment environments based on the organization's objectives:

1. **IT engagement**: As noted in the previous chapter, engagement with IT should start in the PILOT phase. It can be a time-consuming exercise to establish the prerequisites for all required environments and set up these environments in compliance with the policies and practices of the organization. Platform vendors should also be engaged with IT to guide requirements for each environment.

2. **Deployment models**: The PILOT phase is also an ideal time to consider whether these environments will be established on-premises, on the cloud, or as a hybrid environment. The security requirements (firewall, encryption, etc.) will vary based on these deployment models. The deployment model will also drive the roles and responsibilities for BCP/DR, backup, audit, maintenance, upgrades, etc.

3. **Scope and support**: Define the scope of each environment early on. For example, how ML models will be used and supported in development and test environments. These factors will help define each environment's roles, responsibilities, and level of support.

4. **Licensing model and terms**: Each IA platform (core platform and adjacent technology) will have a unique licensing model and terms. Thus, it is essential to ensure that appropriate licenses are available to

set up these environments. In addition, establish governance to manage the life cycle of these licenses from onboarding to retirement.

5. **Access levels**: The access levels for the IA platforms should be established in compliance with the organization's security policies. A process should be set to ensure that access is granted to each environment based on the business needs. The access levels should be audited regularly and revoked timely.

6. **Data security and privacy**: Will mock or real data be used in the test environment? Like the production environment, will test data be encrypted in the test environment? Given the role of CoE, how will PII data be handled in all environments? etc. These factors will help establish data security and privacy policies for each environment.

7. **Controlled environment**: Establish governance to enable features of IA platforms to avoid enabling a feature in the production environment without testing in lower environments. The sandbox environment is an ideal choice to have all features enabled to encourage experimentation.

8. **Code changes process:** Code changes should not be allowed in testing/staging and production environments. Like the software development life cycle, the code should be promoted from development to testing/staging to production. Setting up CI/CD pipeline is highly recommended to ensure this promotion works as designed.

9. **Maintenance schedules**: All these environments require regular maintenance, from security vulnerability patches to IA platform software upgrades. The maintenance schedules should be set in conjunction with IT to facilitate rollback if needed. In addition, a process should be defined to test software upgrades in the lower environments before upgrading the production environment. If you are using a vendor's SaaS deployment environment, set up a cadence for reviewing release notes from the vendor to test timely.

10. **Monitoring**: Like other enterprise applications, set up monitoring and alerting for the IA environments as well. It is commonly set up for the production environment.

11. **Audit**: Like logs of other enterprise applications, archive the bot's logs to your organization's Security Information and Event Management (SIEM) system.

12. **Backups**: Ensure IT has a backup of all the data, which can be restored easily and swiftly in the event of BCP/DR.

DEVELOP OUTLOOK TO SCALE THE IA PROGRAM

It is worth mentioning that, for many organizations and IA leaders, scaling an IA program is an uncharted area with greater complexity and sophistication. The five facets discussed previously have granted us visibility into essential focus areas for scalability. However, there was also an emphasis that the actual complexity and challenges lie in tying all these facets together to make it all work. Thus, IA leaders must develop an outlook to manage these novel complexities and unique challenges to scale the IA program. To help build this outlook, we offer a balanced and holistic viewpoint mapped to the traditional people, process, and technology perspective:

- The people perspective addresses various internal and external partnerships IA leaders should develop.

- The process perspective presents various transformational and leadership activities that offer challenges and must be addressed holistically.

- The technology perspective encourages treating the IA ecosystem as a new application layer on top of the existing business application layer.

Treat IA Ecosystem as a New Application Layer

The IA ecosystem is proliferating rapidly, with various vendors focusing on providing a spectrum of capabilities ranging from general purpose to business function specific. This ecosystem is growing with commonly used components like RPA platforms, IDP platforms, conversational AI components covering chatbots and virtual agents, or a collection of out-of-the-box and custom ML models. This ecosystem is also growing in using unstructured and semi-structured data extraction to advanced analytics. Although not very common yet, there is specialized adoption of Blockchain and IoT technologies with a combination of other IA capabilities.

As illustrated in Figure 7.7, it is of utmost importance to recognize that the IA ecosystem introduces a new application layer on top of existing systems-of-record (CRM, ERP, etc.) and systems-of-engagement (Office Solutions, Messaging, etc.) applications. This layer must be well understood to appreciate the challenges of scaling the IA program. From a security perspective, this additional layer introduces new security risks and challenges that need to be mitigated. From a business continuity point of view, this layer presents an extra point of failure that should be dealt with. Finally, from a support and operations angle, this layer adds the complexity of dealing with IA-specific support needs, which require special skills and dedicated resources.

Security

While expanding the IA program enterprise-wide, the access levels of bots broaden increasingly across business applications, introducing new forms and levels of risks and security challenges. Given that the IA ecosystem involves software and hardware systems, the security risk factors are similar to the traditional cybersecurity risks. However, the IA ecosystem has the potential to introduce novel forms of security risk as it effectively adds a new application layer on top of existing business applications.

This layer presents additional security risks and challenges that must be evaluated and mitigated. At the minimum, we must ensure that the security levels applied to knowledge workers (humans) also apply to digital workers (bots). At the same time, one might argue that these measures must be more restrictive since, unlike human workers, unattended bots can run 24/7, thus extending the window of security risk while introducing a new vector for cyberattacks.

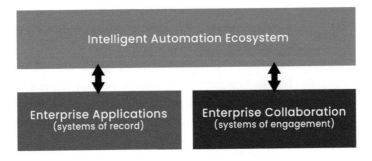

FIGURE 7.7   IA Ecosystem Adds a New Application Layer.

Business Continuity and Disaster Recovery Plan

With the enterprise-wide growth of the IA program, we significantly increase business functions' dependency on the availability of the IA ecosystem. Hence, any prolonged downtime of components of the IA ecosystem could dramatically impact the business processes, especially during business-critical times. Furthermore, the IA ecosystem is a layer on top of the business application layer; thus, this layer's BCP and DR need should be recognized and addressed before scaling.

Support and Operations

The operationalization of bots is paramount for the success of a scalable IA program. However, the full potential of the IA program can only be achieved by ensuring the sustainable support and maintenance of the bots. Therefore, the IA ecosystem must be treated as an independent application layer on top of the business layer that requires the same or better levels of support addressing exceptional SLAs, performing corrective and preventive maintenance activities, and conducting business continuity tests. Moreover, the IA operators must be trained to support the ecosystem requiring specialized skills to monitor, troubleshoot, and maintain AI-enabled bots.

Develop Successful Partnerships

IA leaders must develop and maintain numerous successful partnerships to sustain growth and accelerate the adoption of the IA program. These internal and external partnerships are required regardless of the point in the IA program's journey (see Figure 7.8).

Business Units

A value-driven partnership with business units is essential for the success of the IA program. Each business unit will have its specific needs and priorities; thus, these partnerships must be developed and maintained at the individual business unit level.

Executive sponsorship at C-level is essential to scale beyond a certain degree. Similarly, buy-in and support from the head of a business unit are required to scale within the business unit. Scaling will be expensive; thus, adequate funding will be necessary for an enterprise-level adoption of the IA program.

In addition to sponsorship from the BU head, suitable engagement is a prerequisite with each business unit where IA will be deployed.

FIGURE 7.8 Establish a Holistic, Successful Partnership.

This engagement must be driven by a "Champion" appointed and empowered by the BU head. These engagements can aid in accelerating several activities necessary to run an IA program.

- **Stakeholder engagement:** The business stakeholders' partnership is necessary for sponsorship and funding for the IA program. These stakeholders should also be engaged in defining the business continuity plan for their business area. In addition, the business champions can help establish awareness, change management, and communication plan for their specific business unit.

- **Pipeline management:** For the continuous growth of the IA program, it is essential to have a healthy pipeline of automation opportunities and projects. The business unit's contributions are required to run the discovery workshops to identify and prioritize these opportunities, aligning with their specific objectives.

- **Delivery management:** You will need more profound engagement with business units and functions to deliver projects successfully. The business SMEs should provide test data to build and test the bots and further engage in user acceptance testing. You must ensure appropriate time allocation and resources are provided to conduct this testing successfully.

- **Training & upskilling:** To establish a successful enterprise-wide continuous learning mindset and persona-based training program, you will need this partnership to prepare custom content for the program. This partnership will also be required to prioritize the training program and engage business teams regularly.

Information Technology

A well-established, active, and empowered IT engagement is vital for the success of the IA program. The IT engagement is critical to building a solid foundation for scalable infrastructure and is also required for ongoing business application stability, application upgrades, and testing. This constant engagement is also critical for conducting a detailed technical review to avoid any unexpected load on the business application caused by running bots.

In most organizations, the IA practice starts from a business function and grows into other business units. Thus, these business functions don't have the knowledge and skills to stand up and maintain the hardware infrastructure and software ecosystem needed to run a successful IA program. Therefore, IT engagement is essential to avoid developing a shadow IT in the CoE.

- **Infrastructure:** The partnership with IT will help you drive the hardware infrastructure required to deploy and set up the IA ecosystem. More importantly, this partnership will help you set up the IA ecosystem in compliance with your organization's policies and practices. Further, when the IA program grows, you will need to forecast your hardware infrastructure footprint to ensure that IT is prepared to deliver sufficient infrastructure to scale. Finally, this partnership will be paramount for cloud management activities and planning once you have migrated or adopted an IA ecosystem component on the cloud.

- **Architecture:** The IT enterprise architecture group and IT architect will help you set up suitable architecture for your organization. Since we are building the layer of the IA ecosystem, the IT architects will review this layer, provide feedback, and assist in adapting your organization's best practices. The IT architects will also ensure that you have adequate access and testing practices for the IA ecosystem. Most importantly, the IT architect will also assist in evaluating AI-enabled emerging technologies you are considering for your organization.

- **Application owners:** You will also need to partner with business application owners to help them understand the need for the IA bots using their business applications. They will help you grant required access for the bots, perform reviews of the automation request, and, most importantly, ensure that an adequate level of testing is performed, including load and performance testing necessary for the application.

Information Security and Privacy

As the IA program grows, security and privacy increasingly become a critical concern and potential cause of failure of the complete program. Therefore, a well-established engagement governed by well-defined information security and data privacy policies is required to manage the bot life cycle from granting business application access to execution and, eventually, proper retirement of the bot.

The role of information security and data privacy starts when the IA ecosystem component is being evaluated for adoption. Then, as the IA program grows, information security and data privacy become an even more important partner to ensure that the IA program is aligned with the organization's policies and industry regulations:

- **Security:** The partnership with the security team will help you establish the ongoing need for IAM and PAM for bots and users, and the credential management for bots ensuring your IA program complies with your organization's policies. The security team will also help you establish a process for regular internal and external audits of the IA ecosystem. In the case of security violations, you will need guidance from information security to remediate the issues promptly.

- **Data privacy:** The partnership with the data privacy team is essential to ensure that the IA program complies with your organization's privacy policies and the industry vertical's data privacy regulations. Given that the digital workers will operate like human knowledge workers, they would need the same level of access to the data, potentially exposing these data to developers and support team members, including employees and contractors. Thus, these policies, defined with assistance from the data privacy team, will help set up a process for creating and maintaining the test data with appropriate data masking and encryptions. The privacy team can further help review the solution design of the automated process. The privacy team can also help build custom privacy training modules for employees and contractors, covering the policies and actions required by developers, testers, and support teams.

Vendors and Service Providers

The partnership with vendors and service providers is equally important as the internal partnerships of your organization.

- **Product setup:** You will need help from the vendor of the IA ecosystem component for the proper license management and respective product setup and configuration.

- **Product roadmap:** This is the most crucial aspect of this partnership. You will need to understand the roadmap of product and vendor investments to develop your strategy and roadmap of how to progress with the adoption of various capabilities. The senior leaders from vendors can help you and your business leaders better understand their vision and roadmap of the product. Establish a half-yearly or yearly cadence with the vendor to dive deeper into the roadmap. This might also help you influence the roadmap aligned with your priorities.

- **Product support:** Ongoing product support and incident management will be required. Additionally, if you are maintaining an on-prem environment, then you will also need assistance with software upgrades.

- **Training:** The continuous learning mindset and ongoing training of CoE, business teams, and other teams is required to scale the IA program. The vendors can help provide technical and functional training through on-site/virtual workshops. They can also offer training content through video modules or documents, which can be integrated with the learning management system of your organization.

- **Services:** The rapidly growing IA ecosystem is also becoming complex in functional offerings and technical capabilities. It is impractical for any organization and service provider to catch up and become an expert in all tools, platforms, and practices required for a scalable IA program. While setting up the foundation to scale, you will inevitably hit the situation where you need some external expert's assistance to build the capabilities for your organization. This is the time to tap into some niche service providers to fill in the technical gap of your organization.

Peer Organizations

Peer benchmarking is a powerful way to compare and contrast your journey with other organizations on similar paths. These partnerships can be built with the help of vendors and service providers or by attending forums, customer meetups, and industry conferences. You can also gain general benchmarking information using reports from Gartner, Forrester, Everest, and other consulting organizations.

- **Products and vendor:** This partnership will help you gather relevant success stories to bring to your discussions with business, IT, and security groups. This partnership will also help you understand the challenges and proven remediations in other organizations which might work for you.

- **Best practices:** These partnerships will also help establish best practices for governance, operation, support, and development.

Recognize a Holistic View of Activities

At this point, it is evident that IA leaders will require an adequate technical appreciation to evaluate, select, and evolve the IA ecosystem using emerging technologies. It is also clear that IA leaders must develop several internal and external partnerships to govern and operate a sustainable and scalable IA program successfully. Nonetheless, the recipe for successfully scaling the IA program is by maintaining and managing a holistic view of the transformational activities across the program while focusing on generating value for the organization. IA leaders must be willing and equipped to solve a constant and dynamic puzzle, illustrated in Figure 7.9, based on the organization's priorities and policies.

1. Identify and define the strategic value and desired outcomes that will drive the means for identifying automation and optimization opportunities and the tools necessary for automation.

2. Cultural transformation management activities to develop trust and adoption of IA.

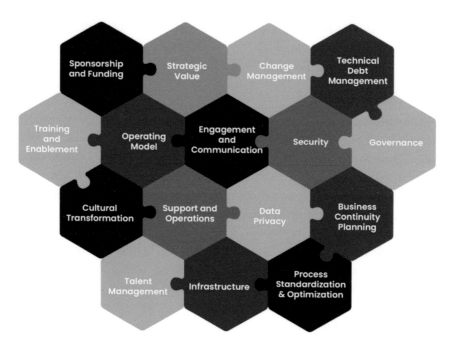

FIGURE 7.9   Holistic View of Transformational Activities.

3. Change management activities to engage and empower employees for broader adoption of IA.

4. Secure the most essential sponsorship and funding required to scale smoothly.

5. Regular governance activities to deliver strategic value for the business.

6. Ongoing operational activities to deliver quality automation on time and within budget.

7. Adopt process improvement practices to reduce process fragmentation using BPM, process mining, and task mining capabilities.

8. Support and operations to monitor, troubleshoot, and maintain bots by a centralized or business-aligned support group. This group manages the activities necessary to ensure that business-critical digital workers perform as expected per the SLA.

9. Institute business continuity planning and management in partnership with business SMEs and IT.

10. Establish a well-defined engagement and communication plan for all planned and unplanned activities.

11. Set up infrastructure management practices to drive the infrastructure's stability, scalability, and security.

12. Establish a security risk mitigation plan adopted throughout the entire life cycle from the conception of the automation idea, captured through documentation of the requirements for the development, testing, deployment, and maintenance of bots.

13. Establish data privacy practices by closely working with the data privacy team.

14. Institute technical debt improvement plan by defining best practices, adopting IT and security policies, decommissioning bots, etc.

15. Talent management by establishing a process to upskill existing employees, define career paths, set up internship programs, and staff augmentation through professional services.

16. Training and enablement activities to establish an upskilling program for stakeholders such as business SMEs and users, business leaders, business analysts, IA developers, IA operators, solution architects, infrastructure engineers, etc.

The following section addresses the holistic view to scale the IA program, covering best practices, challenges, and remediation. IA leaders are encouraged to use this list as a reference and to map out their 360-degree view of transformational activities.

CHALLENGES AND RECOMMENDATIONS

Forrester's report showed that over 50% of organizations with RPA programs couldn't scale beyond ten bots (Joseph & Clair, 2019). Why can some organizations scale their IA program while others struggle? Scaling an IA program is an uncharted area with greater complexity and sophistication. Let's dive deeper into the challenges you might encounter (or should anticipate) and the preparation you may need to remediate these challenges. Note that the list below is not in any order; thus, readers are encouraged to use it as a reference based on their needs.

Identify Strategic Value

Given that the most common adoption of IA technologies starts from a small proof of concept by a business function in a silo, the automation practice starts with bottom-up task-level automation. Although these task-level automations are essential, organizations cannot quickly transition to high-valued processes without further top-down sponsorships and funding.

It is common to see that organizations lack a way to identify and prioritize high-valued business workflows to automate by leveraging IA capabilities. This challenge is driven by various reasons ranging from omnipresent fragmented processes to a lack of optimized and standardized processes.

Forrester Research recently predicted, "Companies with advanced automation programs will obliterate – not merely beat – the competition." Organizations must realize quickly to move away from low-hanging or low-value processes to gain the full potential of IA. Organizations must move beyond ROI to find transformational and strategic value with an enterprise-wide IA strategy. Organizations must institute the strategic value of the IA program aligned with business goals (Clair, 2021).

The strategic value must be set at the organization and the individual business area levels. In addition, IA leaders must identify opportunities to augment business processes by integrating AI-enabled capabilities to unlock long-term strategic business value. **Emilie Ly**, Senior Director, BPM & RPA at VMware Inc., emphasized, "It is essential to connect business transformation goals to automation efforts using key metrics to show the value realization."

Some of the traditional strategic values that organizations have considered are listed below. Each business area should prioritize its combination of these values associated with its business processes.

- Reduce or optimize cost

- Increase revenue growth

- Improve process efficiency

- Improve customer experience

- Reduce regulatory and compliance risks

- Increase employee productivity

- Improve employee satisfaction

Process Fragmentation and Standardization

Non-standard and fragmented processes are the most significant barrier to implementing IA at scale.

Typically, an end-to-end business process involves executing a set of complex steps involving handoffs between business functions and systems, creating fragmented processes. Automating sub-optimized fragmented processes doesn't bring long-term benefits. Thus, optimizing and standardizing the end-to-end process is imperative before automating.

IA leaders must explore and establish practices for discovering, documenting, standardizing, and optimizing business processes. These practices can be established by leveraging a combination of BPM, process mining, and task mining tools and techniques.

1. Business process management (BPM) practices are essential for suitable process discovery. These practices, enabled by the collection of tools, techniques, and technologies, assist in identifying, analyzing, and documenting existing "as-is" business processes. Moreover, these documented end-to-end processes are enormously valuable in identifying opportunities to standardize, optimize, and automate.

2. Process mining can provide valuable process insights to understand how business processes are genuinely working, highlights deviations from the expected process paths, and assists in discovering opportunities to improve the process.

3. Task mining tools and technologies help collect and analyze users' workstation data. Task mining combined with process mining can provide a complete view of how a business process is operated.

The insights collected using BPM, process mining, and task mining can be used to identify process fragments and opportunities to remove process redundancies and standardize process variations.

In the context of scaling IA programs, it's worth noting that the organizations that have achieved tremendous success in implementing and scaling IA enterprise-wide have combined IA implementation with process standardization and optimization. Chapter 2, "The Value of Business Process Discovery," covers this aspect in detail, including the current product's capabilities and everyday use cases.

Sponsorship and Funding

Having an executive sponsor might seem like an obvious ask. Still, when it comes to scaling IA enterprise-wide, its importance is much more significant than having a typical project leader role.

- Sponsors create trust in the IA program.

- Sponsors help drive organizational change management and cultural transformation.

- Sponsors anticipate needs aligned with business objectives and deliver strategic value.

- Sponsors assign the right people to the right roles and empower them to be successful.

- Sponsors and business heads will help secure funding and priority for automation opportunities.

- Sponsors must have a stake in the project's success.

- Sponsors must stay engaged throughout the process to help teams as new challenges arise.

- Sponsors and senior leadership can only navigate specific political challenges.

Cultural Transformation

An organization's culture is one of the most significant barriers to digital transformation. Research shows that up to 70% of digital transformation initiatives fail (Forth et al., 2020). There are many reasons for failure, but it all comes down to the organization's culture.

What cultural challenges are experienced while scaling IA enterprise-wide? Generally, it all comes down to the fear that automation could take my job and that digital workers could replace human workers. IA technologies generate significant apprehension in employees as it has the potential to automate entire job workflow involving both manual and cognitive tasks. This fear results in resistance to change and, in some cases, potentially sabotaging the program.

So, what does this cultural transformation mean in the context of IA? It's a work environment where human and digital workers work side by side (William, 2021). The workforce strategy with an optimal mix of human and digital workers described in the previous chapter (Figure 6.7) is an ideal human and digital workforce integration. The workforce integration strategy mixed with the implementation and maintenance of a humble organizational culture (described in Chapter 5) drives the successful digital transformation of adopting IA enterprise-wide.

In the context of scaling IA enterprise-wide, some essential best practices to implement the cultural transformation successfully are given as follows:

1. The organization's cultural fabric must empower employees to be part of the transformation. Employees should be incentivized to drive the transformation of their job and should be rewarded for their innovation in defining their digital roadmap.

2. Continuous learning and upskilling should encourage employees to take career advancements into their own hands.

3. Executives, digital transformation leaders, and business leaders must actively engage in the cultural transformation process. It is critical in establishing, communicating, and enabling transformational goals. Leadership should have ongoing open and candid conversations about the strategic objectives of human and digital workforce integration.

4. Humble culture must be embedded in the organizational processes, which encourages the behavioral norms – (a) accurate awareness is prompted, (b) competent mistakes are tolerated, (c) transparent and honest communication is rewarded, (d) openness to the ideas of others is valued and modeled, (e) employee development is prioritized, and (f) employee recognition is practiced regularly.

Change Management

Poor or no change management is one of the top barriers to scaling the IA program. Given the "quick wins" nature of the implementation of RPA and IA, change management is often overlooked. Moreover, adopting IA enterprise-wide requires change across all levels of the organization, from people, business processes to technology.

As covered in the previous chapter, at the onset of the adoption of IA, a proactive change management approach must be developed and executed to create awareness and change readiness across the organization. Further, it's essential to recognize that the change management activity will not be a one-time activity; thus, organizations should be prepared and well-invested for the long haul.

The change management program should be targeted to help answer the questions and concerns of employees at many levels.

- I'm not aware of RPA and IA.

- I've heard about RPA and IA, but I'm not fully aware of what they can or can't do.

- Oh! I was not aware that my organization has an IA program.

- I have heard of the IA program, but I'm not aware of what they do.

- I want to contribute to the IA program, but process automation is not the focus of my job.

- I'm aware of the pain points of my business processes. Nevertheless, it's working, so let's not change it.

- I'm concerned that robots will take away my job.

Identify Key Stakeholders

The success of change management crucially depends on the proper identification of stakeholders. These stakeholders are the ones who can influence the change and able to provide ongoing support for the success of the program. They can also become champions or evangelists for the IA program in their business areas.

Launch and Run a Communication Campaign

Launch an internal communication campaign, starting with the change management communication during the adoption of IA and commencement of CoE. It is required to have ongoing engagement and active communication between internal and external stakeholders. Surprisingly, many organizations lack structured engagement and communication among all internal and external stakeholders. A sample communication plan is shown in Table 7.2.

Engage Employees for Continuous Improvement

We can't emphasize enough how crucial a continuous improvement mindset and employee engagement are for the success of the IA program and digital transformation. This plan can address the following:

a. Well-planned and periodic learning sessions are organized through hackathons, brown bags, or lunch-and-learn sessions.

TABLE 7.2   Sample Communication Plan

Communication	Target Audience	Recommended Frequency	Communication Channel
IA Program Newsletter	Enterprise-wide stakeholders	Monthly	Enterprise social network, intranet CoE page, Email
Roadshow/ workshop	Enterprise-wide stakeholders or customs for a business function	Quarterly	In-person and/or virtual meetings
Engagement	Enterprise-wide stakeholders or customs for a business function	Monthly/ quarterly	Brown bags, lunch and learn, Hackathons
Operational Communication	Business SMEs, application owners, IT, etc.	Need basic	Email or status readouts to address project statuses, incidents, upgrades, planned or unplanned downtime, training plans

b. A well-structured feedback loop is received from employees about the IA program, e.g., employee survey after completion of every project or training.

c. Like knowledge workers, digital workers are an integral part of a team and are supervised by functional line managers.

d. Digital workers are assigned names, perhaps based on themes. Communication (emails, etc.) generated by digital workers is signed by their names.

e. Digital workers are an integral part of employee engagement tools, e.g., listing digital workers in the organizational charts in HR systems as direct reports of functional line managers.

Support and Operations

The study conducted by Forrester in 2020, "Barriers and Best Practices for Scaling RPA," reported that 45% of organizations deal with bot breakage every week or more frequently (Business Wire, n.d.; Ariola, 2021). Further, the Robocorp Survey: The State of RPA 2022 found that 69% of survey respondents experience broken bots at least once per week. So regardless of where your organization stands, bot failures are imminent (Robocorp Survey: The State of RPA 2022, n.d.).

Let's first understand why bots fail. This is where the excitement of building bots meets with the ultimate reality. As IA leaders, we must maintain the balance between the two. The last thing we want is to automate the processes on the business end but create support (manual or otherwise) for bot operators, losing the fundamental purpose of automation. This planning becomes even more critical as organizations move to cloud-based services relinquishing the infrastructure control to IA vendors. Figure 7.10 lists some examples of scenarios causing bot failures (note that this list is presented with the intention of helping develop an understanding of possible failures).

Unfortunately, bots' supportability is an afterthought in most organizations. The operational practices used for proof of concept or small-scale deployments will not work with a continuously increasing number of bots. We can't emphasize enough how essential is the operationalization of bots for the success of a scalable IA program. The full potential of the IA program cannot be achieved without ensuring the sustainable quality and maintenance of the bots.

To that end, CoEs should:

a. Establish centralized or business-area-aligned support and operations team.

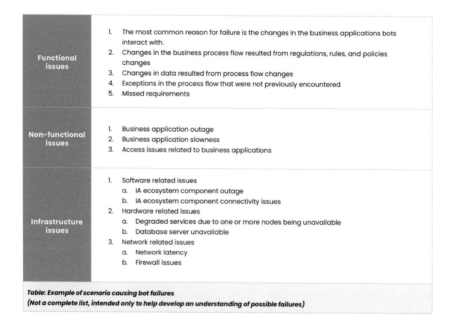

Functional issues	1. The most common reason for failure is the changes in the business applications bots interact with.
	2. Changes in the business process flow resulted from regulations, rules, and policies changes
	3. Changes in data resulted from process flow changes
	4. Exceptions in the process flow that were not previously encountered
	5. Missed requirements
Non-functional issues	1. Business application outage
	2. Business application slowness
	3. Access issues related to business applications
Infrastructure issues	1. Software related issues
	a. IA ecosystem component outage
	b. IA ecosystem component connectivity issues
	2. Hardware related issues
	a. Degraded services due to one or more nodes being unavailable
	b. Database server unavailable
	3. Network related issues
	a. Network latency
	b. Firewall issues

Table: Example of scenario causing bot failures
(Not a complete list, intended only to help develop an understanding of possible failures)

FIGURE 7.10 Example of Scenario Causing Bot Failures. (Not a complete list, intended only to help develop an understanding of possible failures.)

b. Operationalize monitoring, alerting, tracing, troubleshooting, and fixing the issues proactively and reactively.

c. Define operational RACI (responsible, accountable, consulted, and informed) involving CoE, business SMEs, IT, and IA vendors.

d. Establish SLAs with business and IA vendors.

Services Provided by the Production Support Group

In essence, the primary role of this group is to offer and facilitate supportability by ensuring that the bots run as expected per the established SLAs.

Supportability of Bots   Supportability refers to a readiness to monitor, alert, troubleshoot, and maintain bots. The supportability of bots should be thoroughly considered at every step (requirement gathering, development, testing, and deployment) of the development life cycle. E.g., design and development best practices must be followed to ensure sufficient information is available in logs to troubleshoot an incident. The support team is granted suitable access to the business applications, data, and logs to support effectively. Proper access to bots' infrastructure, logs, and transient and final data is also granted to troubleshoot any incident effectively.

Effective supportability practices also establish the proper support levels and SLA with business groups, as listed in Table 7.3. These topics must be well documented and communicated to all stakeholders. It should also address at what stage the vendor's product support team will be involved, how escalation will be managed, and how resolution will be achieved.

TABLE 7.3   Support Levels

Support Levels	Scope
Level 1	Requires general monitoring and basic bot maintenance like scheduling jobs, starting or stopping jobs, and basic configuration maintenance
Level 2	Requires use case understanding, developers' skills to provide adequate solutions or workaround with occasional code changes to resolve the issue
Level 3	Requires deeper functional and technical understanding of the use case and solutions

1. **Stability of bots**: First and foremost, the support team ensures the stability of bots by confirming that all adopted best practices are followed during development and testing. The support team can enforce a checklist of items before releasing the bot code to the production environment.

   Furthermore, the support team will troubleshoot and fix software and configuration errors for unstable bots. Bot operators will require development skills to build solutions or workarounds to resolve these issues.

2. **Incident management**: Capture and manage incidents for production bots. Each incident must be assigned a priority level based on criticality, which will dictate the resolution time. In addition, the support team should have a way to trace and tie the incident with the change causing the incident. E.g., recent code commits in the source code repository are one way to capture changes before the production release.

3. **Bot utilization**: The support team also assists in improving low bot (and license) utilization due to improper and imbalanced distribution of processes. A balanced approach will be required to maintain the isolation of business functions, avoid overlap of data-intensive processes, and manage the uptick in the volume.

4. **Change request management**: The changes are requested for many reasons, including bugs, missed requirements, regulatory changes, data changes for business applications, and minor process changes for the business application. The requested change must be treated and tracked similar to the original project.

5. **Business application stability**: Develop and maintain partnerships with business application owners and IT to maintain regular communication about any changes in the application which could potentially cause bot failures in the production environment. Participate in testing such changes proactively.

6. **Stakeholders communication**: The production support team will perform an array of stakeholder communication and coordination, e.g., the release of process automation, system downtime, process failures, and planned maintenance.

7. **Monitoring and reporting**: Establish how and what to measure and perform well-defined monitoring and reporting covering predefined KPIs using actual data from the production environment. Implement custom reporting for stakeholders across all business areas.

8. **Access management**: Typically, support and operations will have admin-level access to the production environment of the IA ecosystem. As a result, they could assist in managing effective roles-based access management for the complete IA ecosystem.

9. **Auditability and traceability**: Drive auditability and traceability through detailed logging of digital workers' activities and operations performed across components of the IA ecosystem. Establish partnerships with the auditors to develop a suitable mechanism to enable and manage audit requirements.

10. **BCP and DR**: Own and participate in developing a BCP/DR plan and executing periodic mock BCP tests. The support team is vital in triggering and managing BCP during downtime.

Business Continuity Planning/Disaster Recovery (BCP/DR)

Unfortunately, this is the least discussed topic about scalable IA programs. When organizations achieve a level of scale, many business functions and processes are onboarded on the IA ecosystem. These processes could range from a set of simple steps to a complex multi-function workflow involving the cognitive capabilities of the IA ecosystem, resulting in a significant dependency of business functions on the availability of the IA ecosystem. Any small or prolonged downtime could dramatically impact the business process, especially during business-critical times like deal closing for sales or patient record access during critical healthcare processes. The IA ecosystem is a layer on top of the business application layer; thus, the BCP needs of this layer should be recognized and addressed before scaling.

On the one end, intelligent digital workers can effectively contribute to business continuity plans and are seen as the future of business continuity by addressing the need to handle a spike in volumes and staff fluctuations due to vacations, switching jobs, etc. However, on the other end, the outage of these intelligent digital workers at scale could cause a more significant impact on the organization, especially when these digital workers are involved enterprise-wide for critical business processes.

Moreover, the longer the digital worker takes over, the greater the organization's risk of process knowledge loss when human workers are not operating a process. Organizations might need to involve knowledge workers swiftly to take over business processes if these digital workers are out of commission for extended periods. Over the period, transaction volume for these processes could increase, and organizations may not have enough knowledge workers available to manage these processes. Proper knowledge management of automated processes must be documented and maintained, and clear responsibility of business and CoE must be established to maintain business continuity.

Large and matured CoEs have started recognizing the need for BCP. The CoE and business stakeholders are crucial in defining, documenting, and testing a BCP plan for all the business functions involved. Each business function or unit would require its own BCP and DR plans to address its needs and priorities during downtime.

Business Readiness

Business function teams and SMEs should be aware of the business continuity needs, and these steps should be considered during the development life cycle of process automation. In the ideal scenario, the requirements documentation should clarify how BCP will be handled for the business process. In addition, the development and deployment of bots should thoroughly consider the scope of BCP.

I. Contingency plan during bot downtime

Each automated business process requires a contingency plan to be operated when digital workers are unavailable over a certain period.

1. Each business function and area should define criteria about when to trigger the contingency plan.

2. Each business function should establish the process's severity and priority level. When a business function includes multiple automated processes, these severity and priority levels should be relied on to trigger a sequence of BCP actions during downtime.

3. Each process should have a documented contingency plan triggered when digital workers are unavailable beyond SLA.

4. The contingency plan should address whether the business processes will be handled manually, automated, adopt a workaround, or adopt a hybrid approach with a combination of those mentioned above.

5. CoE, the production support team, and business SMEs must be involved in documenting the contingency plan. In addition, the same group will be involved in assessing and executing steps during downtime.

II. **Knowledge management of automated processes**
When the end-to-end process or critical parts of a business process are automated using unattended intelligent digital worker, business function risks losing essential process knowledge when knowledge workers have not manually operated those key steps or used data points to make crucial decisions regularly over a long period.

1. Each automated portion or end-to-end process should be documented under process standard operating procedures (SOP).

2. Most importantly, these steps should be updated and maintained when changes are performed for the respective bot.

3. Driven based on the severity and priority levels of the processes, these steps must be manually tested as part of the mock BCP test.

4. Process audits should verify the coverage and correctness of these steps.

CoE Readiness
The CoE should own the development and execution of the BCP plan in partnership with IT and business SMEs.

1. CoE must be prepared for BCP/DR for the complete IA ecosystem, including components deployed on-premises, on-cloud, or vendor-managed.

2. CoE should have a resourcing and action plan to test and trigger BCP/DR for each business function supported.

3. CoE must perform periodic mock BCP tests for the entire IA ecosystem.

4. CoE must adopt practices to capture the evidence of the BCP test and submission for review and audit purposes. Note that it might also be required for regulatory purposes.

Infrastructure

Typically, the setup of IA CoE commences in a non-IT area of an organization with a proof of concept whereby, generally, some shortcuts are taken on infrastructure setups. Due to ROI-focused practice and rapid delivery commitments, infrastructure stability, scalability, security, and compliance are often overlooked. **Julio Viquez**, Senior Manager, Intelligent Automation at VMware, Inc., confirms, "One of the most overlooked items is establishing platform upgrade and maintenance plans. This could very rapidly add lots of technical debt to the IA ecosystem, which could negatively affect the growth plans."

The acceleration of the IA practice ultimately depends on having a robust and scalable infrastructural foundation. Additionally, the setup and management of IA infrastructure require specialized and scarce expertise covering classic IT infrastructure management, software development, DevOps process, software architecture practices, Machine learning model development and MLOps practices, containerization, and cloud management, to list a few.

The common pitfalls experienced without having a robust infrastructure setup and compliance with IT policies and practices are as follows:

1. Low hardware resources utilization due to improper distribution and allocation

2. Out-of-sync privileges and access rights between development, testing, and production environments

3. Out-of-sync application versions or features between development, testing, and production environments

4. Considerable efforts for upgrading IA platforms

5. Long waiting time to commission bot machines or servers for temporary increased capacity

6. Lack of required backups for DR and BCP

7. Lack of logging needed for compliance audits

Further, the setup and management of the set of IA platforms require an expert understanding of how these platforms interact with enterprise business applications. It requires technical software architect knowledge to ensure the IA ecosystem infrastructure's stability, performance, security, and scalability. The CoE must work in close partnership with IT and drive the requirement of infrastructure and architecture. In some cases, CoE might have to manage it all. However, CoE must avoid becoming "shadow IT" and conduct infrastructure management in close partnership with IT, following the organization's policies and practices.

The architecture and infrastructure, described as one of the facets of scalability, need to be looked at holistically, and infrastructure management and practices must be defined per your organization's business objectives. There isn't a "one-size-fits-all" architecture, and the most optimum approach is to start small, grow the size, and add additional functional components as the need arises.

1. **Controlled growth**: The IA infrastructure can be categorized into components needed to run a common IA platform (such as RPA platform and IDP platform) and the infrastructure required to run the bots. The infrastructure needed to run bots can proliferate depending on the demand and number of projects delivered regularly. Thus, deliberate efforts should be made to have proper and optimized VDI/VMs infrastructure utilization. Its utilization should be monitored and adjusted by rearranging bots on VDIs/VMs, so hardware resource utilization is high. Some RPA vendors also support running multiple bots on the same VDIs/VMs simultaneously using different bot IDs, which should be considered a resource optimization option while maintaining security and compliance.

2. **Burst capacity management**: The infrastructure demand can arise for temporary scenarios like quarter-end and year-end processing, holiday shopping for retail, or one-time bulk data upload/download, requiring many bots to run in parallel. The CoE should be prepared to quickly address such demand and provide the hardware and bots needed. In addition, it's essential to have a process to rightsize the infrastructure after using it for the burst capacity.

3. **Ease and speed of deployment**: The CoE must be prepared with processes and tools that enable the ease and speed of provisioning VDIs/VMs and bots as and when needed. These processes can be implemented using multiple alternatives, from having an end-to-end CI/CD pipeline and self-serviced infrastructure management capabilities for on-prem and cloud infrastructure.

4. **Cloud adoption and deployment**: Most IA platforms are now offering cloud alternatives. In some cases, the SaaS alternative is a preferred deployment mechanism. The CoE must develop expertise, operating, and governance practices to manage the complete infrastructure in the cloud to hybrid deployment.

5. **Multiple environment/tenant support**: As per the Deployment Environments section (herein Chapter 7), it is strongly recommended to set up multiple environments (on-prem) or tenants (cloud) – Production, Dev/Test, and Sandbox. Each environment should be isolated and have its own operating best practices.

6. **Security**: In partnership with IT, the CoE must define a procedure to manage security vulnerability patches for all hardware and version management of all software. Depending on how the hardware is provisioned and operated in your organization, IT may or may not be responsible for the security vulnerability patches for the hardware owned by CoE. Similarly, it will be the responsibility of the CoE to upgrade all IA software components to ensure that the IA ecosystem is up to date in capabilities and feature sets.

7. **BCP and DR**: As covered in the BCP/DR section (herein Chapter 7), COE must establish DR and BCP processes working with IT, information security, and business SMEs. All hardware and software data must be part of a regular backup. The BCP plan should be tested periodically.

8. **Infra for AI/ML models and other platforms**: The deployment of AI/ML models and related platforms could require specialized hardware to train and use the ML models. Therefore, the MLOps practices to deploy, consume, manage, and support the ML models should be considered. The CoE should partner with the data science team to establish and manage this setup.

Security

When organizations expand the IA program enterprise-wide, they gradually widen the access levels of bots across multiple business applications. This introduces new forms of risks and new levels of security challenges. Therefore, special attention must be given to developing robust practices ensuring secured bots and infrastructure are deployed to automate business processes with sensitive data (How to Ensure Robotic Process Automation Security, n.d.).

At the minimum, it is essential to ensure that the security levels applied to knowledge workers (humans) also apply to digital workers (bots). In many cases, these measures must be more restrictive as, unlike human workers, unattended bots can run 24/7, thus extending the window of security risk while introducing a new vector for cyberattacks. At the same time, it is also essential to recognize that if bots are designed, developed, and deployed correctly, they are more secure and less error-prone than humans. Thus, bot security must be considered throughout the entire life cycle, from the conception of the automation idea, captured through documentation of the requirements for the development, testing, deployment, and maintenance of bots.

Security Challenges

In many ways, the security risk factors of IA are like the traditional cybersecurity risks. However, the IA ecosystem has the potential to introduce new forms of security risk. As discussed, the IA ecosystem essentially adds a new application layer to existing business applications. This layer presents additional security risks and challenges that must be acknowledged, evaluated, and mitigated. Further, you will likely have one or more components of the IA ecosystem deployed on the cloud. Or running bots on the cloud to automate business processes on the applications deployed either on the cloud or on-premises. While defining the mitigation plan, the on-prem/on-cloud, SaaS, private/public cloud, hybrid, and multi-cloud deployments must be considered carefully.

For brevity, we will evaluate these security risks under two broad categories. **First**, the risks introduced due to the poor design of bots, and **second,** the general cybersecurity threats experienced with any software systems, including the IA ecosystem.

1. **Poor bot design and development process**

   One of the leading causes of security risk with digital workers is linked to the development process, or lack thereof, concerning poor design, lack of development best practices, and inadequate testing and maintenance. The good news is that we can adopt some well-proven and time-tested techniques from traditional software development practices while still keeping it lightweight. Some examples of poor bot design issues are as follows:

   - **Unsecured data**: Poor design lets sensitive data in the transient or final stage on shared drives.

- **Unauthorized access**: Access is available to bot IDs beyond the need to manage the data for the process.

- **Sensitive data leakage**: Data clean-up is not performed once the digital worker has completed the execution of the process.

- **Unplanned and untested load**: Digital workers can process a larger volume of transactions than human workers in a very short period, creating an unexpected burden on system resources and resulting in degraded service or system outages.

- **Generic bot IDs**: Using generic bot IDs leads to excessively broader access to business applications.

- **IT and InfoSec**: Lack or absence of IT and information security engagement results in non-compliance with the organization's practices and policies.

- **Regulatory non-compliance**: PII data are leaked on shared drives.

2. Cybersecurity threats

- **External malicious actors**: The digital workers (bots) open a new avenue for malicious actors to compromise bots to gain access to business applications and confidential data.

- **Internal malicious actors**: The threat where an inner actor like an employee or contractor gains access to or modifies bots to gain access to confidential data.

- **Software vulnerabilities**: Susceptible to zero-day vulnerabilities of the software components of the IA ecosystem, business software, and operating system environment.

- **Network vulnerabilities**: The threat where a bot exposes a network vulnerability enabling hackers to gain remote access to the network.

Security Risk Mitigation

A well-rounded approach is needed to mitigate security risks associated with the IA ecosystem. To implement solutions that align with the practices and policies of your organization, all the 21 steps listed

Evaluation of IA software systems	1. Conduct a thorough security review of software products of the IA ecosystem. This process typically starts before the procurement and onboarding of IA software products.
Establish Business Process requirement practices	2. Business process assessment should consider the security requirements and determine whether the security team review is needed before automation. Prioritize automation only when security concerns are addressed 3. IT and Information security reviews must be performed when onboarding a new business function or application onto the IA ecosystem. Engage with business application owners to plan for load and performance testing to prevent DDOS attacks.
Establish Architecture best practices	4. Establish and implement role-based access controls 5. Establish Identification and authentication 6. Use Privileged Access Management (PAM) tools 7. Assign a unique ID to each Bot 8. Use Password vault to secure passwords 9. Establish a plan for security vulnerability patches 10. Use Multi-factor authentication (2-factor authentication) where possible
Establish development and testing best practices	11. Conduct thorough code reviews (both automated and manual) 12. Testing using mock data 13. Implement data encryption for production data 14. Conduct Performance and Load testing
Support and Operations practices	15. Establish Data security practices for support 16. Archive bots and IA platform logs, use these logs for security and observability platforms 17. Establish Monitoring practices 18. Establish Auditing practices 19. Conduct periodic audits and risk assessments which could lead to retiring digital workers in time. 20. Establish a business continuity plan and conduct periodic tests
Expand Training scope	21. Introduce specialized data security training aligned to usage of bots and IA ecosystem

FIGURE 7.11   Steps for Security Risk Mitigation.

in Figure 7.11 must be thoroughly assessed and implemented in partnership with IT, information security, data privacy experts, and business SMEs.

Data Privacy

Although data privacy is integral to data security practices, it requires special attention since digital workers directly deal with sensitive data of customers, employees, processes, transactions, etc. (How to Ensure Robotic Process Automation Security, n.d.).

Privacy Data Management

To remediate concerns related to the handling of PII data, privacy data management practices should be established to answer the following topics. These practices must align with the rules and policies of your organization and thus should be developed in partnership with data privacy, information security, IT, and business SMEs.

1. How should CoE manage access to PII data? For example, do we need to mask data for development and testing purposes?

2. Do we need to encrypt data in the process execution's transient and final stages?

3. When and with whom can the data be shared?

4. Do developers and operations need additional custom training to support data privacy scenarios?

5. When should the data privacy team be involved in reviewing the process changes and implementation?

6. What data privacy controls are required for citizen-developed automation?

Development Life Cycle

Data privacy concerns should be addressed at various stages, from process evaluation, requirement gathering, development and testing, support, and operations.

During process requirement gathering, a procedure should be in place to capture if PII data are involved in the automation steps and whether they need to be evaluated by data privacy experts. These requirements should also address how PII data can be accessed and managed in various environments like development, testing, and production.

The development practices should ensure that mock data, instead of real PII data, are used during development and testing. The code review should ensure that sensitive data are not exposed through logs, configurations, and process files. The data are encrypted in the production environment, and the support team should have access only on a need basis to troubleshoot an issue.

The periodic audit process must include a review of these measures to ensure that any procedural violation can be addressed in a timely fashion. In addition, the systematic risk assessment should take extra steps to evaluate bots leveraging PII data to ascertain if these bots can be decommissioned if not in use anymore.

Technical Debt

Technical debt is a term often used in the context of software development and IT deployments. Technical debt typically results when quality is sacrificed for speed of delivery and needs future rework or consistent maintenance. Unfortunately, technical debt is one of the least talk-about challenges when considering the scale of IA programs.

Automation is also susceptible to technical debt, which could result in expensive ongoing maintenance. Poor design and development could cause chronic maintenance issues leading to downtime, roadblocks to scaling, and even negatively impacting ROI.

In the context of the IA ecosystem, the technical debt can build over the period through bot code, IA infrastructure, and integration of IA ecosystem components. For example, suppose the bot code is not upgraded periodically to be compatible with the rest of the IA ecosystem. In that case, it might work for an extended period due to backward compatibility but will certainly hurt scaling efforts at some point. For infrastructure, if the bots are deployed on too many different shapes and sizes of configurations of VMs or VDIs, it will impact the speed of scaling until standardization is performed. Finally, if nonstandard practices are used to integrate IA ecosystem components, it will affect the scaling pace.

In essence, continuous and balanced improvement through modernization (newer features and technologies) and standardization (coding best practices, software integration, hardware configurations) are required to prevent the buildup of technical debt.

Factors Influencing Technical Debt

1. **Lack of best practices**: Missing coding practices in the early stages of the journey and introducing them later will create significant technical debt when keeping up with modifications in best practices. Further, the ongoing demand for security, data privacy, and support will need regular improvement of best practices and hence the generation of technical debt.

2. **Bot's failures**: "Quick fix" for broken bots to resolve failures in the production environment is one of the most common causes of generating technical debt.

3. **Software compatibility**: Ideally, the related component of the IA ecosystem should be on the same compatible version. Keeping them on out-of-sync versions might work for a certain period, but it is bound to fail or cause issues slowing the pace of scaling.

4. **Non-standard infrastructure**: It could apply to any component of the IA ecosystem, but it has the most noticeable impact on VMs and VDIs of the robot farm.

5. **Non-standard integration of IA ecosystem components**: Integrating using custom APIs when standard APIs are available.

6. **Infrastructure upgrade**: Infrastructure upgrade is the perfect time to pay back technical debt. However, it could also generate some debt if shortcuts are taken.

7. **Security updates**: Regular mandatory security updates mixed with non-standard infrastructure could create unmanageable combinations of software and hardware.

8. **Decommissioned bots**: Bots are deployed and active even when not used for an extended period.

9. **ML models**: AI/ML models bring another level of technical debt if not maintained and re-trained regularly.

Steps to Manage Technical Debt

Adoptions of following best practices can help minimize technical debt:

1. Establish coding and configuration best practices.

2. Establish an engagement plan with IT and security to keep up with rules and policy changes.

3. Define a process for ongoing maintenance, including IA ecosystem software and hardware upgrades, maintenance of ML models, migration of key infrastructure components to adopt the latest policies (e.g., Internet Explorer discontinued), business application upgrades, and regulatory and compliance changes.

4. Define a process to decommission bots not in use through periodic audits and reviews.

5. Develop standardization for infrastructure to ensure that limited combinations of hardware configurations are in use. Leveraging OS-level golden image or containerization technologies can greatly help here.

6. Institute bots monitoring and reporting to capture bot's utilization helping in maintaining a well-distributed and utilized infrastructure.

Talent Gap

One of the top barriers to scaling IA programs is the scarcity of qualified talent. Unfortunately, IA talent is not only hard to find, but it also takes time and effort to develop, and it's even harder to retain them.

A qualified IA talent requires significant technical expertise involving RPA, AI/ML, and other emerging technologies. They are also expected to have the business acumen to understand complex business workflows and transform business problems into effective IA solutions. Additionally, they are expected to be a change agent to be able to champion the change empathetically.

Evidently, the most substantial demand for AI/ML talent has come from non-IT departments like marketing, sales, customer service, finance, and research and development. These business units leverage their skills for customer churn modeling, customer profitability analysis, customer segmentation, cross-sell and upsell recommendations, demand planning, and risk management. These talents are often hired directly into these departments with clear use cases in mind to learn the intricacies of the specific business area and remain close to the deployment and consumption of the models they create.

Factors Influencing the Talent Gap

Many factors are still driving this talent gap. IA leaders must recognize the impact of this gap and prepare their organizations to deal with it.

1. **New and specialized skill development:** On the technical front, IA requires unique and specialized skills with expertise in technologies like RPA, AI/ML, NLP, IDP, conversational AI, virtual agents, Smart Workflows, and many more. Limited professionals have gained sufficient experience in the collection of these technologies. Finding someone with RPA or AI/ML background might be manageable, but finding someone with expertise in holistic IA technologies is more challenging. On the non-technical front, IA requires specialized business domain knowledge or acumen to identify and discover IA opportunities, which can be done successfully only when one has a great deal of knowledge about the functional capabilities and limitations of IA technologies. These functional and technical combinations are challenging to develop and retain.

2. **The rapidly growing spectrum of emerging technologies:** The rapid growth of the IA field and associated emerging technologies

exacerbate the issue as the demand for new skills continues to emerge. While organizations are struggling with successfully implementing complex automation involving document processing and unstructured data, industry analysts, vendors, and niche organizations are attempting to integrate Blockchain or Quantum computing into the mix. It's not practical for learners to catch up with the pace of growth in the IA field. Digital transformation leaders must prioritize the needs of their organization and accordingly develop and hire talent to address the current requirements.

3. **Not much clarity in the industry about the career path:** Individual IA roles are well defined, and it is clear to professionals what it takes to become a developer, architect, solution engineer, business analyst, and project manager. However, it is unclear to professionals and even organizations what career paths employees can take and how they can develop their careers within the organization. In many organizations where IA initiatives started with a pilot in the last few years, a business (or sometimes IT) leader had to initially carve out this role for themselves as a side gig or an enhanced role. Compared with traditional software development or business operations roles, most organizations have not yet defined a clear career path for IA careers. This is a discouraging factor for aspirants interested in a long-term career in the IA field.

4. **Lack of specialized courses in colleges and universities:** Current undergraduate and graduate students are unaware of the IA field in most colleges and universities. This is even true for Technology and Business majors like Information Systems, Computer Science, Business, and Management. In the last few years, some colleges have started partnership programs directly with IA vendors to teach and train students of specific majors to better prepare for careers in automation. Although it is a great start, specialized IA courses are still far from being incorporated into the college curriculum. Additionally, some of these partnership courses are limited to the hands-on training on the vendors' platform and may not cover the nuances of establishing, managing, and running a CoE to achieve the full potential and enterprise-wide scale of the IA program. This book is an attempt to address such needs directly.

Actions Are Needed to Fill in the Talent Gap

When scaling the IA program, very specialized skills are needed to develop, maintain, manage, and sustain the growth and quality of the program while preserving ROI. A multi-faced approach to filling the talent gap needs to be examined. Each organization must evaluate its need for IA talent by looking at the variety of sources available to them to fill in the gap.

1. **Upskill existing employees:** Due to budget limitations, hiring new skills may not be a sustainable solution. Overall, it is cost-effective to train and upskill existing employees. A methodical approach, described in the next section of training and upskilling, is needed to establish and run an enterprise-wide training and upskilling program. Minimally, every organization inclined to have a scaled IA program must:

   a. Identify skills required for the IA program based on the organization's priorities.

   b. Secure dedicated budget and resources for learning and development.

   c. Run custom training and upskilling program to close the talent gap.

   d. Start early and iterate quickly to align with industry evolution.

2. **Establish clear career paths:** Business and digital transformation leaders must develop clear career paths in their organizations. These career paths must support the organization's functional, technical, supporting, and governing roles.

   1. Process-oriented career paths (Lean Sigma to Process and Task Mining)

   2. Technical career paths

   3. Leadership career paths

   4. Business domain career paths

3. **Staff augmentation for specialized roles:** To scale the IA program, specialized technical skills may not be available in your organization. Thus, external partners should be considered to have these resources available on relatively short notice and with less hassle for hiring. In many cases, it might require multiple partners to build the foundation and scale further. It is not uncommon to have multiple partners at a certain level of scalability. For example, some partners may focus on large-scale implementation or change management, while others can assist with specialized skills like AI/ML integration, cloud deployments, and testing automation.

4. **Vendor-based professional services:** The professional services groups at vendors are very well versed in their products and can be an excellent asset for implementing complex and niche solutions. These groups can also fill in the talent gap on relatively short notice.

5. **Specialized system integrators to accelerate scaling plans:** As it is clear by now, scaling the IA program enterprise-wide requires a technical understanding, strategic framework, and dedicated resources to accomplish the objective. Some external parties can help you with that and fill in the talent gap you might be experiencing in your organization.

6. **Grow internship and new college graduate pool:** Organizations must consider hiring and training interns and recent college graduates for the long-term benefits and a sustainable talent pool. Investing in training interns and new college graduates is also a powerful cultural signal of how much the organization values its IA program.

Training and Upskilling

Digital transformation is a journey rooted in the need for a quick turnaround to business demands. Organizations must have a continuous training and upskilling plan to prepare for such a turnaround. Specifically, with the constant and rapid advancement in the field of IA, the practitioners involved must have up-to-date knowledge of the capabilities and limitations of IA technologies. To realize the actual value of these technologies, organizations need to have a disciplined approach for continuous upskilling of the employees, requiring a regular investment.

According to the Upskilling for Shared Prosperity Insight Report by the World Economic Forum, upskilling could lead to the net creation of 5.3 million new jobs by 2030, whereby human labor is increasingly complemented and augmented than replaced by new technology. COVID-19 has accelerated this need further, forcing digitalization and automation at a more rapid pace. Although the report highlights the demand at the macro level, a clear parallel is applicable at an organizational level and thus needs to be addressed immediately. Note that digital transformation for an organization depends on many internal and external factors; therefore, the process varies for every organization. Therefore, your organization will need a tailored upskilling program to cater to your specific needs and priorities (World Economic Forum 2021 Upskilling for Shared Prosperity REPORT, 2021).

These training and upskilling plans must be driven based on various functional and technical roles across the CoE and business community. As the IA program grows in your organization, you will need more business users who are well-versed in IA technology and can translate these capabilities into business use cases in a meaningful way. Similarly, on the technical front, to scale IA enterprise-wide, the CoE is required to be at the forefront of technological growth. The CoE must keep an eye on the growing market of IA to be aware of emerging technologies and new vendors that can be leveraged to automate and optimize business processes to gain the full potential of IA across the enterprise. This is not a small undertaking. It would require organizations to periodically redefine and realign the upskilling of the employees involved in the IA program. IA leaders should establish a focused and structured program to train and upskill their functional and technical knowledge workers to be prepared and empowered to contribute to the success of the IA program.

This training and upskilling program should also address the talent gap your organization might be experiencing, and equally important, the program is appropriately mapped with market trends of IA. More importantly, this program should allow existing employees to develop new skills and enhance existing skills preparing them for career development opportunities within the organization. Instead of hiring new people, it's cost-effective for your organization to invest in your existing employees as they are well versed with your business processes, policies and compliance, and the organization's culture.

It is worth noting and reiterating that the success of the IA program relies on the careful implementation of process automation, AI, and ML capabilities to optimize and automate end-to-end business process workflows. With adequate training, employees can be critical contributors to accelerating innovation in your organization. More importantly, in addition to delivering effective IA, well-trained employees can also champion its benefits across the organization resulting in organization-wide culture to adopt and scale IA practices.

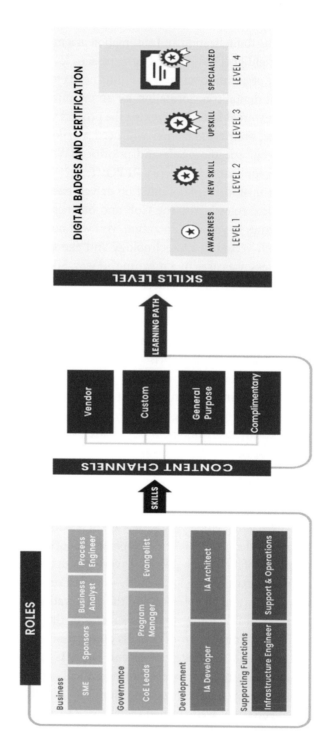

FIGURE 7.12 Process for Establishing an Upskilling Program.

Establish Upskilling Program

To establish the upskilling program in your organization (refer Figure 7.12), start with the end in mind and follow incremental steps to get to the goal. Develop the program's scope upfront and acknowledge that it is meant for the long haul. Note that this program is set up and run by the CoE. The CoE should drive the training framework and scope for the program to ensure that your organization's best practices and adequate coverage of emerging technologies are built into the training modules.

Planning and preparation: First, determine the following aspects of the program and set up guidelines aligned with the organization's priorities and culture:

1. Determine the functional and technical roles in your organization that are required to undergo training and upskilling.

2. Determine how you are going to get the content for training and upskilling.

3. Determine what kind of training content is suitable for your organization. For example, is vendor-provided content sufficient for you? Do you need custom content?

4. Determine what levels of training for each role you need. For example, do you need only an awareness level, or do you need expert-level content?

5. Determine learning paths for each role tailored to the organization's needs and priorities.

6. Recognize gamification that helps develop and strengthen the organization's culture.

7. Engage with HR, Change Management, PMO, and Learning teams to align the training program with the organization's policies and culture.

8. Select a learning management system to incorporate learning paths and gamification.

9. Prepare for a complimentary custom mechanism for learners through workshops, hackathons, office hours, brown bags, or lunch-and-learn sessions.

1. **Establish training roles:** The upskilling program needs to be based on various personas or roles. List and categorize these functional and technical roles in your organization, covering the roles from

CoE and the business community across the organization. Typically, these roles will fall under one of the following categories.

a. **Business roles**: This will cover roles like business SMEs , business analysts, process engineers, sponsors, or other business stakeholders.

b. **Governance roles**: This will cover a group of roles required to run the CoE or IA program, for example, CoE leads for centralized or federated CoE, project or program managers, champions, or evangelists who would advocate the IA program throughout the organization.

c. **Development roles**: This will cover technical roles needed to develop IA automation, ranging from IA developer, ML engineer, data scientist, and IA architect.

d. **Supporting roles**: This will cover a group of supporting roles like infrastructure engineer, support, and operations.

2. **Define skills:** Once the roles are established, define each role's required and expected skills. These skills definitions could range from bullet points to formal job descriptions.

3. **Categorize training content:** Categorize the training as awareness, enablement, and specialized for each role. This categorization will help drive the difficulty level of training content and help establish the skill levels for gamification.

4. **Collect training content:** Collect and prepare content relevant to each skill level. There are multiple sources from where the training content can be collected.

a. **Vendor**: Many vendors have product-specific training modules, generally online modules. Some vendors also provide hands-on virtual or on-site training programs.

b. **Custom**: Custom content for your governance, operating, and support models will be required. This will cover the best practices and policies of your organization.

c. **General purpose courses**: You may also want to consider some general-purpose or specialized online courses available at LinkedIn Learning, Coursera, edX, and other Massive Open Online Course (MOOC) platforms.

d. **Complimentary**: Note that eLearning modules will not be sufficient for certain roles. Establish a complementary mechanism for learners through workshops, hackathons, office hours, brown bags, or lunch-and-learn sessions.

5. **Define learning paths:** Define learning paths for each role. These learning paths will be a collection of training modules required to complete a skill level. These may include modules from all sources like vendor, custom, and general-purpose courses. These paths must be communicated to the participants with well-defined effort levels and time expectations. Progress tracking and reporting must also be established for each learning path.

6. **Set up training program using LMS:** Set up the program using an LMS of choice in your organization. Various Sharable Content Object Reference Model (SCORM)-based LMSs allow integrating training modules from vendors and other sources alongside the organization's custom training modules. These LMS systems typically assist in:

a. Creating custom training modules.

b. Collecting and integrating training modules from vendors and other online sources.

c. Delivering the content aligned with your organization's brand and culture.

d. Tracking and reporting the progress of the participants.

e. Gamification through the issuance of digital badges and certifications.

7. **Set up gamification:** Adopt gamification by issuing digital badges and certificates. Have custom certification for each skill level or include vendor or industry-specific certifications.

8. **Run training program:** Organizations can adopt any suitable approach to run the program, e.g., they can first target all roles and set up only awareness level of training for these roles, or they can plan for upskilling for only one role and set up all skill levels and

learning paths for this role. We can't emphasize enough that dedicated resources and funding will be required to run a successful enterprise-wide training program.

SUMMARY

In the last few years, IA programs have exhibited the potential of IA technologies to deliver cost reduction, risk reduction, and revenue growth. As a result, business and digital transformation leaders are now considering IA as a critical part of their strategy. This adoption will grow with more complex automation and unprecedented technology integrations, and organizations should prepare themselves for such a scale of change.

Business and digital transformation leaders must determine what strategy would work for their organization. Working with executive sponsors and business leaders, organizations must evaluate the strategic value of IA as a differentiator for their business. Executive sponsors and business leaders must assign adequate resourcing and sufficient funding to set the program for success. Each business area must consider standardization and optimization of the processes before automating. Leaders must enable and empower employees to become change agents. Finally, organizations must invest in talent development and upskilling. All this while ensuring governance that leads to the scalable, agile, and secured foundation for scaling.

The IA leaders must recognize that the actual complexity and challenges for scaling the IA program lie in tying various independent facets together to make it all work. Therefore, IA leaders should maintain a complete view across all scalability aspects while focusing on generating value for the organization. It's worth repeating that to implement and scale the IA program, IA leaders must maintain a holistic view at multiple levels – (a) the IA ecosystem consisting of an ever-growing technology footprint, (b) internal and external relationships required to deliver the transformation successfully, and (c) preparation for managing the everyday challenges and activities.

If done right, overcoming the challenges and barriers to scale, organizations in nearly every industry and market sector can reduce costs, increase productivity, improve customer experience, streamline operations, and grow revenue by leveraging the full potential of IA.

Enjoy automating!

CHAPTER 8

Intelligent Automation Unleashes Diversity and New Ways of Learning

Aftab Ahmed

INTRODUCTION

As has been covered in earlier chapters, IA is not merely a piece of software or a tool for building solutions. Instead, IA represents a new way to think. A new way to think about processes and workflows and when activities should take place. A new way to think about what humans should be doing and where their efforts bring the most value to an enterprise. And it represents a new way to think about how to teach and learn skills that will be needed in the 21st century.

In this chapter, we explore the journeys being taken by enterprises as they seek how to best train and tool their employees. The authors feature the new methods that are being discovered and adopted in the pursuit of delivering mass learning, but in a customized way, with the use of bot-a-thons (also sometimes termed hack-a-thons). These innovative and agile learning techniques seek to provide rapid, experiential, applied learning, overcoming many of the shortcomings of traditional learning techniques.

DOI: 10.1201/9781003276128-8

Bot-a-Thons Present a New Way of Learning

A few years ago, no one had heard of the term bot-a-thon. Now, bot-a-thons and similarly termed hack-a-thons have entered the lexicon as the newest way to teach practical technology skills.

How It Works

While the precise structure of a bot-a-thon will vary according to the specifics of the organization running the activity, the basic premise is to gather a group of people who want to build IA solutions for problems that they bring forward to be solved and pair with experienced developers that can help them to build the solution and teach them on the job. Participants are usually tasked with completing some training beforehand, typically from the extensive online materials provided by IA vendors and other third parties, which usually only requires a commitment of a couple of hours. Then, typically, a few prioritized problems are worked on in small pods of 3–5 people, with participants wishing to learn the IA tool being led by an experienced developer on a specific problem being solved. By learning through experience in a networked fashion, with the immediate satisfaction of solving real problems, bot-a-thons present a whole new way of learning. See Figure 8.1 for an illustration of a typical bot-a-thon structure.

Bring Your Own Problem to Be Solved

One of the biggest problems with most traditional training experiences is the lack of being directly applicable to your specific needs. With any

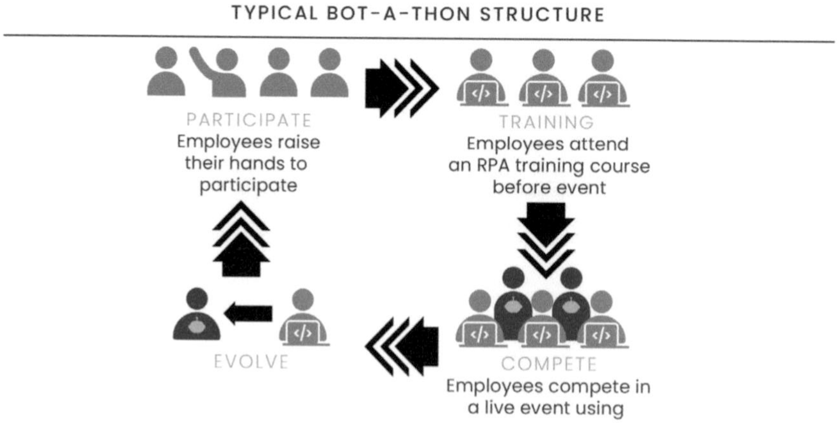

TYPICAL BOT–A–THON STRUCTURE

PARTICIPATE
Employees raise
their hands to
participate

TRAINING
Employees attend
an RPA training course
before event

EVOLVE

COMPETE
Employees compete in
a live event using

FIGURE 8.1   Typical Bot-a-Thon Structure.

form of mass training, tailoring to meet the exact needs of all participants is unrealistic. Thus materials usually cover generic examples that convey the widest breadth of learnings, but often at superficial levels. Many post-training surveys capture comments like "not really relevant to what I do" or "hard to apply to my needs." Indeed, most companies acknowledge this in their training philosophy, adopting principles like 70/20/10 for 70% on-the-job experience, 20% interactions with others, and 10% formal classroom or web-based training (Training Industry, Inc, 2020). A bot-a-thon is centered around solving *real problems* that you are trying to solve *right now*. As such, it brings direct relevance and, thus, a sense of ownership to the learning experience that gives the feel of a customized approach.

Learn from Experts on *Your* Solution

As the problem being worked on is personal to your needs, the guidance you receive from the experienced developers collaborating with you in the bot-a-thon provides direct advice on how to build to best meet your process requirements. So, the best practices for working on that particular website interface, that specific business application, or those exact types of document files can be taught and understood as related to your problem. And, of course, those deep learnings can still be taken and applied to similar parameters in future problems you may look to solve.

Collaborate, Network, and Compete with Peers

One of the most important aspects of training is the exposure to others with similar training needs and the opportunities for group learning, collaborating, and networking that it provides. The bot-a-thon concept places all participants on a peer level, as users of the solutions learn together as they discover tool functionality and how to apply it to their solution needs. These network connections are often the starting point for long-term collaboration and support as the bot-a-thon participants form their custom networks for the future development of their IA skills. This is the approach often used with new graduate hires, where initial training provides opportunities for new recruits to bond and build a network to help guide and support them in their future careers. The same principle, just with a new skill set. By being among peers, bot-a-thons also allow for the introduction of gamification, where there is often a scoring or prize element to the training event for the best or most inventive solutions to be recognized. In the history of humanity, competition has often led to the greatest advances in capabilities, and bot-a-thons seek to exploit this natural human emotion to further the learning experience.

Immediate Application of the Learning to the Real-Life Use Case of Value to You

Another failure of traditional learning techniques is the lack of a direct line to how the training is useful to you. Think of learning calculus in school and wondering when you would ever need to use that in real life! Bot-a-thons provide for the human need for instant gratification by working to deliver a usable solution at the end of the training so that the participant can realize immediate value. Avoiding hypotheticals, fake scenarios, or simplified setups helps make the benefits of the training and the tool's capabilities feel much more real and, thus, more beneficial.

Rapid Experimentation, Learning from Failures, and Celebrating Successes to Build Confidence

As illustrated in Figure 8.2, a key evolution in IT project methodology in recent years has been the shift in sequential, so-called waterfall methodology that goes from point A to point B to the use of "agile" practices of define, design, test, and repeat in a series of iterations that builds up a solution incrementally. This provides several benefits, including faster and more adaptable framing of business needs, quicker delivery of partial solutions, more real-time and frequent feedback from users, and an ability to experiment, trial, and improve rapidly. In addition, failures become just learning experiences rather than disasters, and successes can be built on and celebrated to grow the participant's confidence in their skills.

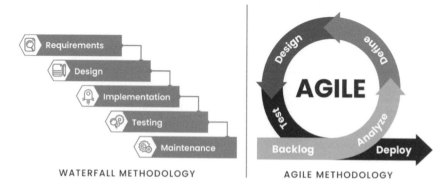

FIGURE 8.2   Waterfall vs. Agile Methodology. (Agile vs. Waterfall: What Is the Difference? Which Is Best for You? 2021).

In summary, bot-a-thons combine the benefits of instructor-led training in a group setting for networking and collaborating, with the benefits of DIY learning and customized experience, all within a psychologically safe yet competitive environment to take people out of their comfort zones and strive for new skills.

A Scalable Approach to Growing and Diversifying Footprint

As has been referenced multiple times in this book, delivering IA at scale is how organizations can realize game-changing results. To deliver results at scale also means educating the organization at scale, either as developers or users of IA. The end user focus of these tools provides such mechanisms when it comes to building that education and related skill sets.

Training Can Be Delivered to Any Size Audience in Any Location

The portability of IA and the proliferation of free vendor and other third-party training make learning IA available to any size audience, from sole learners to groups of hundreds. These training materials are also available extensively online in multiple languages, meaning that geography is also no longer a barrier. This ability to reach out across the full sphere of an organization promotes true diversity and inclusion in ways rarely seen before, if ever, in the technology space for business (Figure 8.3).

Training can be tailored to specific needs, experiences, competencies, processes, business areas, etc.

By combining the benefits of multi-level online training resources and hands-on experiential learning through bot-a-thons, IA training can be orchestrated to deliver a highly customized experience for learners at scale. Online materials typically cover all skill levels, from beginner to intermediate to advanced, providing technical guides on how to use IA

FIGURE 8.3 Popular RPA Training Modules, UiPath Academy. (Learning RPA – Automation Courses | UiPath, n.d.)

tools. The process agnosticism of IA then allows for bot-a-thons to match these learnings with real-world solution needs that account for the business areas and process needs of specific automation opportunities. Bot-a-thons can be targeted to specific business functional areas, geographic groups, tool experience levels, applications involved in the automations, etc., to allow maximum opportunity for aligned solution building and/or aligned learning as best meets the needs of the organization.

Provides Organic Growth Opportunity for Train-the-Trainer Build-Up of Experts

For many organizations, the cost of offering training can be a prohibitive limiter to the scope, scale, frequency, and quality of training offered to their people. Even the largest and most affluent of organizations have to determine how to balance internal dedicated training resources with external training sources with on-the-job training expectations of staff and supervisors. Increasingly, organizations look to leverage online resources and train-the-trainer models to deliver training at an affordable scale. IA's training idiom is completely focused on this approach. Employees start with online training and then learn from experts in partnership while building solutions and experiencing success and failure, leading to a pattern of success from which to build confidence and grow skills. These newly trained participants then become the functional/group/process IA subject matter experts (SMEs) or champions, who are tasked with training their colleagues through the same methodology (see Figure 8.4). With the speed and flexibility of bot-a-thons, a network of trainees can become trainers, providing organic expansion of skill sets and awareness throughout the organization.

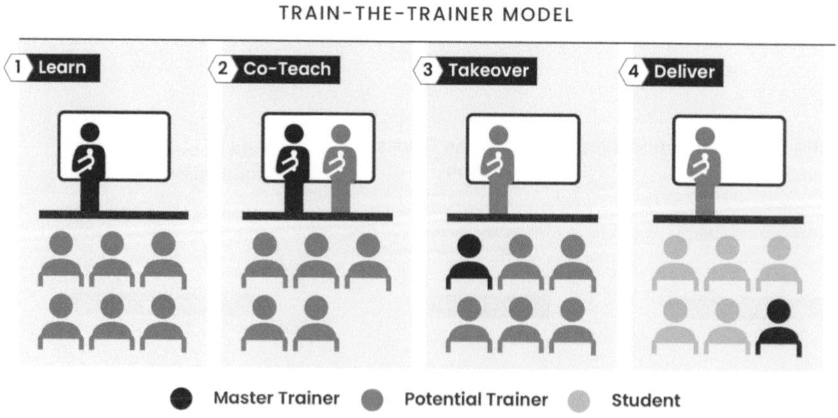

TRAIN-THE-TRAINER MODEL

FIGURE 8.4   Train-the-Trainer Model, Expert Program Management.

Proven Value from Small Learning Investment Promotes Ongoing Investment to Harden Skills

One of the reasons that formal training is relatively limited in most organizations is the challenge of proving the value of such training versus the organic on-the-job training that provides the day-to-day skills that workers need. As businesses, processes, and roles change and evolve over time, maintaining formal training to keep current with all business needs is also challenging and costly. IA bot-a-thon training methodology is designed to start very small, building that first set of solutions with a handful of experts and first movers on the citizen development front. By building solutions from the outset, designed to be built and implemented quickly, returns on this small investment can be more readily measured. Users are rewarded with improvements to take to their day-to-day jobs and imbued with curiosity and access to see what else these tools and other emerging digital technology can automate. This, in turn, promotes ambition to learn more, growing and hardening their skills, to solve more complex automation needs, which usually offer even more value and benefits to the organization (Figure 8.5).

Builds Connections Vertically and Horizontally to Foster a Community of Developers at a Pace

A key facet of IA, heavily promoted by the vendors, is the community of practice that has been developed for users of their tools. The major IA providers all boast a network of developers using their products numbering hundreds of thousands and growing by the day. This external setup can also be replicated within organizations, allowing different business areas,

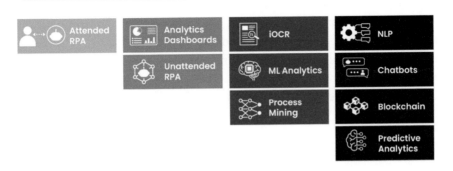

RPA IA SKILLSET EXPANSION EXAMPLE

FIGURE 8.5   IA Skill Set Expansion Example.

functional groups, and geographies to build connections vertically and horizontally, crossing organizational lines and grade-level hierarchy. The experiences shared through building your first solutions and working on shared opportunities allow bot-a-thons to foster a community of IA developers that self-teaches, self-improves, and self-expands at a rapid pace.

The approaches to teaching and learning IA are modeled around building a community of practice that share the common goal of learning how to use IA to improve business processes. These inherently afford opportunities for scaled growth of the IA footprint across the full diversity of the organization, which over time, will see users grow their digital technology skill sets as the tools continue to mature and evolve.

The Huge Diversity of the Population That Can Build Bots

By focusing on the end user and how they work in their day-to-day interactions with business processes, IA is accessible to anyone, anywhere. Indeed, the major vendors have targeted their offerings from a wide range of end users, from kids through STEM development to seasoned developers with decades of experience across many programming languages. This impressive range of clients truly demonstrates how IA is for everyone.

Demographics Are No Longer a Barrier

History has shown that a multitude of demographics is typically at play in each individual's progress through life. Skill sets at work are no stranger to these forces, where a person's age, gender, location, language, education, and financial status all play a huge role in the success of that person. IA is all about accessibility, equality, and empowerment. Skills can be learned as early as when kids first start playing with computers and can continue to be developed throughout an entire career. Location irrelevance is becoming inaccessibly accepted as the new norm in a post-pandemic world. The "apprenticeship" model of IA training with bot-a-thons and alike means that graduate and postgraduate academics are not required to succeed. And with free trial licenses and online training, you don't need to invest money to learn how to use it: a dare-to-try mentality and a willingness to learn become more the determinants of success.

Visual Learning, "On the Job"

As discussed earlier, on-the-job training is the go-to approach for most organizations; typically, 70%+ of all learning is expected this way. The user-interface-focused nature of IA, coupled with experiential bot-a-thon teaching, is a condensed version of this same on-the-job approach. But rather than learning over the full course of time in a particular role, this visual working on real problems, with trial and improvement agile building of solutions, looks to bring all the benefits of on-the-job learning in a fraction of the time, allowing for ongoing skills buildup over time.

Not Just for IT!

A key message promoted by IA vendors is that their tools are for all business users, not just the IT professionals who are typically charged with IT solutions for business problems. Indeed, the non-IT professional is perhaps even more so the primary target customer of these vendors. They have built their tools to look familiar, employing the windows, icons, menus, pointers graphical user interface (WIMPS GUI) that all computer users are familiar with. Click-and-drop technology allows workflows to be executed and connected without needing to write code—connectors to many of the world's most common applications and systems. The message is clear; they believe anyone can build solutions with their software. That's not to say there aren't situations where IT professionals should be involved. Indeed, in governance, maintenance, and support, IT is strongly encouraged to be fully engaged and lead the adoption of IA. But those skills can be focused on more complex solutions, hardening code, error handling, looking for common components for building code libraries, etc. So, pay for IT dollars where you need them most and where they provide the most value.

Keep Technical Competence Close to Business Capability

A common challenge with deploying IT resources in organizations is balancing centralization with localization, managing risks, delivering services, and being cost competitive. Oftentimes, this means business users see IT as a black box that is reluctantly reached out to for solutions (or more often fixes), with little ability to direct or monitor progress or even know if a solution will be coming. Many IT models are evolving to follow a product structure, with IT staff embedded or partnered into business capabilities that support the business structure. However, such a transition is typically slow and full of change management challenges as entities grapple with legacy systems and infrastructure, looking to modernize to today's real-time, cloud-based, gig business. IA offers the ability to build an army of technical competence within end business users, keeping the development of the solutions they need within their sphere of control. This has obvious benefits for pace, cost, and flexibility – all traits that are highly valued in solutions for business needs.

The diversity of people that can be engaged in IA is truly phenomenal. Smart partnering with IT can allow for the maximum exploitation of the potential across your organization to alleviate constraints, reduce costs, increase empowerment, and ultimately build solutions that deliver value and provide a return on experience that benefits all.

CONCLUSION

The flexibility and agility of IA solutions can be mirrored in the training and development of RPA practitioners. Organizations deploying IA technology are embracing innovative learning techniques, customizing on-the-job training to the needs and experience levels of their employees through the use of bot-a-thons to offer experience-focused, real-world-driven, problem-solving training. This allows for rapid development and scaling of RPA practitioners to advance in unison with the IA adoption journey being undertaken by the organization.

CHAPTER 9

Automation Is Here to Stay

Marie Myers

HP Inc.

I N LESS THAN A decade, robotic process automation (RPA) established itself as a legitimate career path and a software technology that is here to stay. When COVID-19 came along, RPA was already growing as an important business strategy, but the pandemic accelerated its deployment globally at warp speed (Roe, 2021).

It has been an amazing transformation, seeing the industry develop at an incredible pace during the last 2 years – much faster than any of us could have ever anticipated. As the industry evolves, we have seen an unprecedented mass exit, "Great Resignation," from the workforce spurred by the pandemic. However, more and more new use cases regarding the "Great Resignation" draw attention to other factors that played a role in the turbulence provoked by the pandemic. Harvard Business Review (HBR) noted five more ongoing factors: retirement, reshuffling, relocation, reluctance, and reconsideration (The Great Resignation Didn't Start with the Pandemic, 2022).

No matter how you look at it, the world market is struggling. With the rapid acceleration of both digital transformation during the height of the pandemic and, most recently, the acceleration of RPA through the arrival of the "Great Resignation," automation is one of the few options to help insulate against employee turnover and loss of key skills.

DOI: 10.1201/9781003276128-9

This chapter explores the most recent trends and provides a view into the potential future of **automation**.

Intelligent automation (IA) has firmly established itself as a technology chief information officers (CIOs) consider an essential part of their toolset, and general managers have deployed this capability throughout their businesses.

Paul Walsh, Head of Digital Technology and Operations at Sony, states:

> That IA has been one of the most critical capabilities in his technology toolbox. Paul says that without IA, most CIOs and business leaders simply would not have been able to advance digital transformation at the pace necessitated. Digital transformation requires a comprehensive suite of tools and capabilities, and IA fits in perfectly.

As seen in previous chapters, the growth of IA as a technology has been accompanied by an explosion of skills in this space. Today, numerous jobs call for specific skill sets. These involve everything from basic RPA developers to more complex jobs requiring a thorough understanding of automation in process mining.

I remember about 7 years ago, I attended a parent–teacher conference with my husband. The school principal announced that we needed to prepare our kids for jobs that did not even exist. I whispered to my husband that I was busy hiring a bot controller and that this was a great example of jobs that were being created.

Today, RPA roles are mainstream. If you take a quick look at LinkedIn, you will see many jobs associated with RPA as a profession. In fact, I did a quick search as I was busy writing this, and thousands of roles specifically require this type of experience. Moreover, job postings ask for experience in platforms from different vendors. This no doubt confirms the incredible demand for these types of skills. As businesses continue to automate, I expect this trend to persist and grow significantly.

In this chapter, I will highlight how the pandemic has not only accelerated this trend but also quite literally established the maturity of technology and the industry today.

The pandemic has undoubtedly been instrumental in rapidly accelerating the world of intelligent automation. At the same time, the industry also has matured to support a clear new segment. The pandemic challenged companies to find new and different ways of working as employees were forced to work remotely, within a very short time, and in a hybrid environment.

Employers suddenly found themselves in situations where in-person collaboration disappeared, and companies' normal controls in specific organizations, such as call centers, disappeared. IA to the rescue! IA enabled call center employees to apply automation principles to their roles and manage their jobs remotely – effectively and more efficiently. Implementing IA would have been much slower and more difficult. Many could not imagine that call center employees did not need to be centrally located to perform their jobs effectively. Today, in a post-pandemic environment, call centers have become acceptable and frankly more effective to embrace hybrid work styles as the new normal. In support of hybrid workstyles, customer service call centers are deploying IA to standardize and speed up agent work to serve customers better, transforming how call centers operate. In addition, many tasks humans originally performed have now been automated and are performed by software bots or augmented by bots enabling hybrid work to take off.

The pandemic also significantly impacted call centers, both in terms of the nature of the work itself, driving volumes higher and resulting in more activity. This increase, combined with the remote working environment, was a key accelerator in adopting IA in this space during the pandemic. Many companies had no choice but to adapt to demand changes quickly. And one of the quickest ways to solve these issues was through automation. This resulted in perhaps the most radical transformation of call center capability in a short period of time. The use and integration of bots into the process is a clear example of how automation, which has become the backbone of customer communication strategy, can benefit a process and enable a much more efficient and effective outcome for customers. For example, let us consider using interactive voice response (IVR), which is defined as "a technology that allows telephone users to interact with a computer-operated telephone system through the use of voice and dual-tone multi-frequency signaling (DTMF) tones input with a keypad (Wikipedia Contributors, 2023)." Using IVR, a bot can provide customer support 24/7 through machine learning, artificial intelligence (AI), and natural language.

According to a study by Markets and Markets:

> The global Call Center AI Market size is to grow from USD 1.6 billion in 2022 to USD 4.1 billion by 2027 at a Compound Annual Growth Rate (CAGR) of 21.3% during the forecast period. The major factors

driving the growth of the Call Center AI Market are the advent of AI by organizations to offer enhanced customer support services is driving growth of the call center AI market.

MARKETS AND MARKETS (2022)

Undoubtedly, this industry will only continue to grow and flourish as more people acquire training in these areas and companies continue to deploy automation technologies.

The most recent advancements in Generative AI will change the ecosystem of how we interact with all software. Generative AI, driven by the recent surge or, should I say, the explosion of interest and use case of Chat GPT, highlights the warp speed at which this entire space is moving. Chat GPT literally burst into our lives and has changed the way we work, code, study, write and communicate. It was astonishing to observe the rapid adoption of this Bot, even in my own family. My children were quick to adapt and use it in their daily lives, and even my husband, who is not always an early adopter of any technology, decided it was definitely something he would use too.

Today the conversations in Boardrooms are focused on what AI outcomes were driven this week, not this year. In fact, I was in a meeting just recently – where some senior executives commented that very shortly, we will be examining the daily, if not weekly, impacts that AI is having on society. In the past, we saw technologies move in increments; in retrospect, it might have been more akin to the pace of a tortoise.

This is why we need to partner with higher education institutions to prepare students for these roles. Many universities are just beginning to acknowledge the importance of automation and IA in their curriculums. Yet, today, very few academic institutions have the vision to prepare students to work with IA adequately. Frankly, greater investments in this area are needed – this is the future. I hope this book will inspire business and higher education leaders to consider adding courses or certificates to complement their degree programs. It is critical that business and academia foster a relationship to ensure students understand automation and possess the necessary skill sets required when they graduate.

Several RPA vendors have worked hard to build relationships with higher education institutions and offer students support through academic alliances. In addition, these vendors are working with universities and workforce development organizations to better equip students and professionals with the skills they need for the automated world we are in today.

One such program to drive interest at the university level has been through bot-a-thons. As a result, bot-a-thon competitions have become more prevalent and allow students to get exposure to the technology and understand the array of applications for various business challenges.

During the bot-a-thon events, students compete against each other to solve a business problem by programming a bot. In addition, there is usually a prize or acknowledgment to encourage participation. While this is a great approach to stimulate interest, there is definitely a need for a more long-term and strategic view of education. This is required to ensure students gain essential coding skills and broader business and analytical problem-solving skills.

During the pandemic period, IA is being leveraged to adapt to the rapidly evolving scenario from an automated collation of COVID-19 patient data for the government to employee health tracking to expedite financial support requests.

The volume of data that needed to be processed could not be handled by humans alone, and IA enabled data around COVID-19 to be tracked more efficiently. As a result, "RPA has emerged stronger out of the pandemic," says **Raghu Subramanian**, founder of Actyv.ai. Raghu was formerly an executive in a leading RPA company, responsible for driving growth in the APJ market.

Raghu has a unique perspective as one of the original employees of a key player in the RPA space and helped develop the business in India from the ground up over the last few years.

Having seen multiple successful use cases of RPA and the transformational benefits that it was delivering to organizations, he was always confident that RPA would be a successful industry.

He notes that while industries are adopting RPA across domains, the maximum use cases and adoption of RPA have been in the Banking, Financial, Services and Insurance (BFSI) industry because it is fairly rules-driven and highly regulated, thereby calling for repeatable process executions.

Raghu says he has seen a lot of preliminary use cases around in the areas of data extraction, migration, and reconciliation. Now he sees how RPA transforms domain-specific processes like invoice processing in finance, know your customer (KYC), compliance adherence in banks, claims registration and processing in the insurance industry, or improving first call resolution and productivity problems in the contact center business."

IA today has the potential to transform most industries where there are rules-based and defined processes. With the rise of new use cases of

IA, there are many new entrants into the sector during the pandemic. The growth of IA has been astonishing. Today, the industry is proliferating with new and emerging startups, including niche players such as Conexiom, that have grown rapidly during this time.

Conexiom and other companies saw opportunities to move into the space and harden up areas of RPA that were once left only for the larger players in areas such as Accounts Payable. Conexiom specializes in financial applications of automation and combines this with process mining to create an end-to-end management system for key financial processes.

For example, **Ray Grady**, former CEO of Conexiom, focused on applying automation principles in the area of financial processing and has partnered with process mining company Celonis to build an end-to-end capability around accounts payable.

As Ray commented, "this is the future of automation, and we will continue to see these technologies converge and provide deeper insights into business operations for today's financial professionals."

Today, he states CFOs and other financial practitioners want a one-stop-shop for their technology solutions, and integrating these technologies is crucial, he says. In addition, they simply do not want to deal with an array of applications and want to see benefits and ROI right up front as quickly as possible.

RPA is not only easy and relatively inexpensive to implement, but it also integrates with other capabilities very nicely, such as process mining.

In most global corporations, turnover or retention in the United States is already running in double digits and is projected to grow further over the coming months. This is creating an enormous opportunity to automate roles that require repetitive actions. This not only takes out the mundane work but also affords the opportunity to de-risk the workforce. Automation is potentially one of the key concepts to solving the high turnover and providing stability to most corporations.

The growth of the entire IA space also spawned considerable executive talent. In addition to those previously mentioned in this chapter, I have had the opportunity to meet with two industry leaders, both of whom worked for key players in the space and have now moved on to other roles and started new ventures.

Both have subsequently gone on to create other fascinating and interesting technologies. A great point to note is that an emerging school of talent is being created, which is part of the greater halo impact of IA.

One of those leaders is **Adrian Jones**, *CEO personar.ai*, who has extensive experience in the IA space. He joined the industry when IA was fairly new and had human-centric connectivity, and frankly, the future of the world was shifting to a digital perspective with human capital.

These technologies augment human capability and help people do their jobs more effectively. What used to take hours or minutes now takes a few seconds. There is no doubt automation is getting faster and improving productivity.

But there remains a residual fear factor surrounding these technologies, similar to the fear surrounding the internet 30-plus years ago.

Technology, says Adrian, is one of those things that changes frequently and will continue to evolve much faster than we probably think. Whether human empathy can keep up is still an issue.

He gave an example of a bank where customers could automate a loan document and get approval on their loan within seconds. The problem was that everyone who got approved was happy, but those who were not approved were disappointed. The process actually went too fast and did not consider the human aspects. We cannot lose sight of that in the future of automation.

Adrian also noted that IA is just one piece of the puzzle. While RPA will continue to be the main workhorse for automated processes, you need other technologies, including machine learning (ML), AI, etc., to make it more efficient and realize its full potential. But the speed of ramping up can be slow, and return on investment (ROI) diminishes when the implementation doesn't happen quickly.

CONCLUSION

There is no doubt the future of the IA industry is bright and will continue to gather momentum in the years to come. This is evidenced by the size of the IA market and the incredible growth that it has experienced.

This is a strong testimony to the size of the market and the kind of growth possibilities for the future. However, others conclude the market will be even bigger due to broad automation adoption across industries.

The recent pandemic has only accelerated the pace of technology adoption, provided an incredible platform for growth, and enabled a broader collection of technologies to grow around it. Combined with the pace of digital transformation, we should see this technology revolutionize professions.

In fact, I recently attended a financial function in Australia, which included a number of CFOs and Board Members. It was astonishing to hear a leader in the accounting field speak to the transformation currently underway in that industry. Nowadays, digital technologies transform accountancy, and IA has automated many elements of the profession, such as testing for audit efficiency.

I am convinced that, in the not-too-distant future, automation and digital skills will be essential for any accountant and should be included in the curriculum for finance students, just as it is critical today to understand how to use Excel.

I am indeed intrigued and excited about the future and look forward to a more automated enterprise that allows humans to focus on value-added and interesting work and leave the more mundane, routine-bound work to the bots.

Moreover, while writing this book, new challenges arose in how we work. For example, several countries are pushing hard to shorten workdays and reduce work weeks by 1 or 2 days. However, with the current gap in skills in the global economy and an increasing desire for more leisure time, few options will enable this change.

Companies and lawmakers across the globe are working to become more agile and flexible drivers of business workweek changes to aid in reducing employee stress without affecting productivity. Recently, Belgium announced workers could opt to work four 10-hour days. In addition, dozens of companies across the USA and UK are testing a 4-day week. California lawmakers are considering a bill to make the standard workweek 32 hours instead of 40 hours. If it becomes law, it will apply to workers in companies with 500 or more employees (CBS News, 2022).

IA has an enormous role to play in the future of work and in potentially providing more meaningful, creative, and strategic work while allowing us to regain balance in our lives.

While automation has the potential to transform how we work truly, it needs to be simplified – low code or no code – so just about anyone can use it and drive it. The internet used to be highly complex; now, it is user-friendly. Automation needs to be the same in the future.

We have come a long way from the early days of RPA when employees feared losing their jobs due to automation. Now we are at the stage where RPA is augmenting rather than replacing humans.

In addition, the speed of advancement of AI, particularly the use of Chat GPT combined with RPA, will create a much more expansive use case. This will undoubtedly affect all technologies and potentially create a double-edged sword for traditional Automation. However, one thing that is certain, is that the speed of transformation will accelerate and create new opportunities for jobs and will present new risks that we will need to navigate as these capabilities are unleased.

I look forward to this evolution and hope it creates a better balance for the world's future and revolutionizes work as we know it.

CHAPTER 10

Periscope into the Future

Shail Khiyara

Turbotic

L EGACY SYSTEMS ARE HOLDING us back from achieving greater technological advances due to their inflexibility and sometimes lack of scalability. Therefore, organizations must move away from legacy systems and
toward newer, more advanced solutions to keep up with the ever-changing
technology landscape.

In an era where carbon (humans) and silicon (computers) unite, organizations must embrace creating a culture that leverages an indispensable
symbiosis of humans and computers and moves us from legacy into the
future. This symbiosis concept states that the two can work together in a
mutually beneficial relationship. For example, computers can assist humans
in complex tasks and provide access to large amounts of data, while humans
can provide the knowledge, intuition, and creativity that computers lack.

The carbon–silicon symbiosis is an important design principle inherent
to the core business and academia partnership mentioned in this book.
The design principle to consider is whether technology should be designed
to add to human intelligence rather than replace it. And the Bizademia
(business + academia) partnership is core to such a symbiosis.

Process automation technologies have revolutionized the way humans
work, providing improved accuracy and precision, higher efficiency and
productivity, and greater control over processes. However, the current
sprawl of multiple technologies that makes up intelligent automation

DOI: 10.1201/9781003276128-10

creates an unnecessary technical debt for many organizations that must integrate these set of technologies on their dime. Often, an unintegrated set of technologies handicaps the automation use case pipeline, as does the notion that – automation is "easy" to deploy and can have an immediate return on investment.

As we look to the future, multiple key trends hover on the horizon, as depicted in Figure 10.1. And, as of the publishing of this book, we have OpenAI's ChatGPT in the mainstream, already making waves with 1 million users in just 5 days. The progress of generative AI has been astounding over the past few years, with significant advancements in natural language processing, computer vision, and creativity. Groundbreaking architectures such as GPT-4 and its predecessors have revolutionized the field, enabling artificial intelligence (AI) to engage in rich, contextual conversations and produce human-like text, music, and images. These state-of-the-art models have shown remarkable versatility across domains, including content

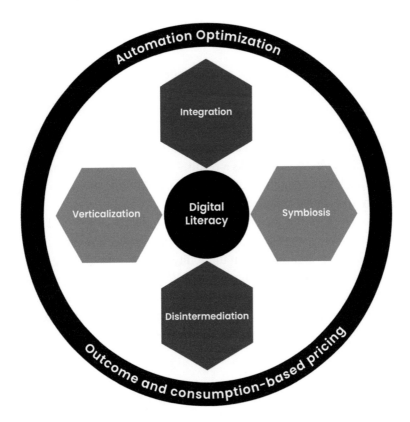

FIGURE 10.1   Key Automation Trends.

generation, translation, customer support, and even aiding in scientific research. Researchers continue to explore the ethical implications and societal impacts of this rapidly advancing technology while working on techniques to reduce the computational resources required and improve the control over AI-generated content. As the field matures, we can expect even more significant leaps in the capabilities and applications of generative AI, with the potential to redefine industries and revolutionize our daily lives.

INTEGRATION

Automation started with a promise – the promise of integrating disparate systems together. However, this unfettered promise has not been realized as automation has grown into a sprawl of technologies. Instead, organizations are seeking a single platform that can unite not just the technologies but the workflows, data sets, and more to help improve their total cost of ownership and, most importantly, the management of automations.

Integrating software technologies can help an organization increase efficiency, accuracy, and flexibility. Today's digital customer expects a seamless experience:

> Process automation brings much-needed efficiencies, but enterprise DNA is essential for the long-term strategy, ensuring business resiliency and innovation. Whether biological or technological, the power of DNA is unavoidable. Today, society is reckoning with how "technological genetics" dictate outcomes, such as how *algorithms impact democracy.* As a result, I have been researching and thinking more deeply about how platform architectures can lead to intended or unintended consequences.
>
> The DNA of the integration approach (you could also call this the API-first approach, as some vendors have) is not a replication of what has been done before but the ability to weave disparate systems together to achieve new things entirely. While other categories make existing processes better, faster, and cheaper, integration platforms beckon companies to don their chef hat and mix up a collection of platforms to see what new recipe they can achieve.
>
> KHIYARA (2022)[1]

On the periphery of intelligent automation (IA) technology hand-offs lies the incredible value. A value that holds the promise of unleashing previously inaccessible data lost in hand-offs between unintegrated technologies. Such debilitating muscle spasms can often weaken an automation program. The future holds promise for further consolidation of such disparate technologies and the emergence of master orchestration or process orchestration solutions that enable the smooth operation of an automation program.

DISINTERMEDIATION

Over the course of the last 2 years, we have seen an incessant consolidation within the IA market. At the same time, adjacencies such as process mining and intelligent document processing have flourished while consolidating at a faster pace than robotic process automation (RPA). Amid this, organizations with core businesses in ERP, IT Ops, and integration platforms have all entered a "check the box" mode of acquiring (not building) automation solutions.

Per *Gartner*, 50% of CEOs and 69% of boards are demanding accelerated growth and operational excellence. However, IA is not the only path to get there. Chapter 6 highlights several building blocks of IA technologies.

intelligent Business Process Management (iBPM) solutions also model, analyze, and optimize end-to-end business processes. And several iBPM players, who are adept at case management, now have RPA functionality through acquisition (Schroeder, 2019).

Organizations with core business around IT Ops see the opportunity to provide RPA through acquisition. Several iPaaS (Intelligent Platform as a Service) companies are also providing automation capabilities. Where RPA or IA was seen as the intermediator between technologies, the future will see further disintermediation of automation as it gets subsumed into other, more mature technologies.

As is the norm with technology cycles, this disintermediation can potentially lead to lower prices and further commoditization of basic automation and other adjacent technologies, creating a more efficient market for software and services.

VERTICALIZATION

The verticalization of software technologies refers to the trend of software companies moving up the value chain and offering more integrated, end-to-end solutions to their customers. RPA and IA, to date, have been offered as horizontal solutions. However, there are several factors that point to a need for verticalization, including the increasing complexity of business processes, the rise of digital platforms, and the growing importance of data analytics and other advanced technologies, including AI.

The key benefit of verticalization is that it allows software companies to offer more comprehensive solutions to their customer, covering all aspects of a business process or workflow. IA, when commoditized, will need to rise above the disintermediation trend mentioned above. One key component to achieving this is identifying, training, and integrating the right domain expertise. Domain expertise provides a deep understanding of the specific industry or business process that is being automated. It is equally important to expand into new markets by developing solutions tailored to that market's specific needs.

As of the writing of this book, we have seen healthcare-specific RPA business (raising close to $1 billion) and credit union-specific RPA business, and this trend will accelerate in the near future.

SYMBIOSIS

Exponential technologies such as ML, blockchain, natural language processing, RPA, and low code will fundamentally disrupt the way we work. It is quite common for industries to combine technologies and obtain a factor ten improvement in, for example, lead time, quality, and customer experience. All of these are possible with the combination of RPA and blockchain.

Blockchain is a collaborative, immutable ledger that offers a distributed, shared ledger among multiple parties to process data and information. One way that RPA and blockchain can work together is in the context of supply chain management. RPA can automate and streamline processes in managing the supply chain, while blockchain could be used to secure a decentralized ledger of transactions within the supply chain. This can help increase the visibility and traceability of goods within the supply chain and help reduce the risk of fraud and errors.

The combination of these two technologies strengthens each by addressing the weaknesses in the blockchain on standardization, scalability, and interoperability and those in RPA around lack of strategic adoption and getting beyond automating human labor. The benefits of this symbiosis will positively impact the supply chain, financial services, government, healthcare, and many more.

DIGITAL LITERACY

Digital literacy will become a broad topic. Digital literacy includes the skills required to facilitate the use of technology safely, effectively, and reliably. Next to human dignity, digital literacy is the greatest gift that can be imparted to every global citizen and is imperative to communicate, find jobs, get a comprehensive education, and socialize. This is where the partnership between academia and business is ever so crucial.

The World Economic Forum predicts that by 2030, more than 1 billion people will need reskilling because their role might have dramatically changed from where it is today (WEF, 2022). Many organizations are building programs to drive digital literacy, and today's best practices might lead to a standardized framework to close the gaps in digital maturity. It is often said that digital transformation is not about technology but about people (Rogers, 2016). Digital literacy needs to be synced up with the company's vision and in the context of the role of the person going through the training, preferably at their own pace – to enable them to be conduits of change. Such a program starts at the top, with the organization's leadership setting the appropriate example. COEs capturing immense knowledge around the automation life cycle should be plugged into an organization's overarching digital literacy program.

The future holds a promise where digital literacy is part of the school curriculum, certification programs, self-assessments, performance-based assessments, and more. Moreover, you cannot do digital transformation without digital literacy.

AUTOMATION OPTIMIZATION

Automation optimization is a category that brings together end-to-end automation orchestration and management capabilities to cut cost, increase performance, and measure business impact – across multiple automation tools. The limitations of current orchestrators and the broad adoption of a multi-vendor approach within organizations are the catalyst for the rise of automation optimization:

> RPA tools have evolved from simple bots that automate single, micro tasks or activities to more complex end-to-end, unattended solutions that can automate entire processes and deliver unprecedented benefits. However, automation management is the automation that automation vendors forgot.
>
> At the heart of RPA is the orchestrator. While orchestrators have improved, and some have moved to become cloud-based, several have not been rearchitected. As a result, the design debt built over the years (with a focus on selling bots vs. managing them) has limited orchestrators to basic operational bot metrics.

The high cost of orchestrators prevents organizations from fully capitalizing on the current limited benefits and integration challenges, making it even harder to incorporate multi-vendor orchestrators into the tech stack. The swivel chair approach (input data from one system to another) that automation is supposed to eliminate is back, with precious resources swiveling to manage multiple automation vendors, extract metrics only to populate Excel and PowerPoints, meticulously and manually maintain bot schedules and reschedules, and painfully pray that bot failure may be detected early.

KHIYARA (2022)[1]

One of the key benefits of automation optimization tools is their ability to automatically identify and manage underutilized or idle automation resources. This can help businesses save money by only paying for resources they need. Another benefit is their ability to provide real-time visibility and control over automation resources, detect errors and downtime, find causes and remediate them, and remove the fragility associated with automation. Automation optimization tools provide a single pane of glass for you to be able to manage your entire end-to-end automation life cycle.

OUTCOME-BASED AND CONSUMPTION-BASED PRICING

There has been a resurgence of interest in outcome-based pricing in recent years, particularly in the pre- and post-pandemic era. In addition, with the rise of digital technologies and the increasing availability of data, it has become easier for businesses to track and measure the outcomes of their products or services produced and to use that information to develop pricing models based on the value they deliver to the customer.

Outcome-based pricing is a pricing model in which the price of a product or service is based on the results or outcomes it produces for the customer. This approach to pricing has become increasingly popular, and one of the key advantages is that it aligns the interests of the customer and the seller. For example, in a market that has pushed the automation risk onto the customer, outcome-based pricing is a desirable path to reduce this risk and ensure that customers are only paying for the value they receive.

Consumption-based pricing, also known as pay-as-you-go or usage-based pricing, enables customers to pay for what they use or consume.

This is the fast track to becoming many businesses' pricing models of choice. It provides agility, flexibility, and the ability to throttle your costs based on the needs of the business. In addition, in an automation environment, where there are oversold bots and not as many processes to automate, plus a collection of technologies to choose from, customers tend to desire a usage-based pricing model.

The future of automation technologies is bright and requires organizations to embrace the creation of a culture and create a symbiosis of humans and computers. Organizations that operate with the premise that technology should be designed to add to human intelligence rather than replace it and make digital literacy the backbone of their business will drive success and distinct competitive advantage. Bizademia (business + academia) partnership is essential to promote literacy and navigate through the current sprawl of multiple technologies and avoid unnecessary technical debt. The inevitable trends of integration, verticalization, disintermediation, and symbiosis, combined with the rise of a focus on digital literacy, will serve the future digital customer with a seamless experience.

CONCLUSION

In conclusion, the rise of IA, bolstered by generative AI, has revolutionized how organizations approach their business processes and the future of work. Bridging the gap between corporate and academia has opened new opportunities for growth and innovation. Advancements in automation technology and the integration of generative AI will actively shape the future of work as organizations adapt and embrace these innovations. However, it is crucial to recognize that more than technology is needed to yield success in IA.

A humble culture that encourages collaboration, diversity, and inclusivity is the key to unlocking the full potential of IA in organizations. As we look towards the future, we expect to see even more advancements in IA, with generative AI driving creative problem-solving and facilitating innovative solutions across various domains. The race to IA has just begun, and organizations must be proactive in embracing the technology to stay ahead of the curve.

Organizations will actively shape the future of work by leveraging the power of IA to create a more diverse, inclusive, and equitable workplace. The IA revolution is here to stay, and it is up to us to harness its potential for the betterment of society and the world.

NOTE

1 https://www.cio.com/article/308690/the-unfulfilled-promise-of-automation-dna-matters.html

Acronyms

AGI	Artificial General Intelligence
AI	Automation Intelligence
ANI	Artificial Narrow Intelligence
API	Application Programming Interfaces
ASI	Artificial Super Intelligence
BCP	Business Continuity Planning
BI	Business Intelligence
BPA	Business Process Automation
BPD	Business Process Discovery
BPM	Business Process Management
BPO	Business Process Outsourcing
BU	Business Unit
CapEx	Capital Expenses
CD	Continuous Deployment
CI	Continuous Integration
CoE	Centre of Excellence
CPU	Central Processing Unit
CRM	Customer Relationship Management
DR	Disaster Recovery
DTMF	Dual-Tone Multi-Frequency Signaling
E2E	End-to-End
ERP	Enterprise Resource Planning
FTE	Full-Time Equivalent
GDP	Gross Domestic Product
GPU	Graphics Processing Unit
GUI	Graphical User Interface
HCM	Human Capital Management
HPC	High-Performance Computers
HR	Human Resource
IA	Intelligent Automation
IAM	Identity and Access Management
iBPM	intelligent Business Process Management
ICR	Intelligent Character Recognition

IDE	Integrated Development Environment
IDP	Intelligent Document Processing
iOS	iPhone Operating System
IoT	Internet of Things
iPaaS	Intelligent Platform as a Service
IT	Information Technology
ITSM	IT Service Management
IVR	Interactive Voice Response
KPI	Key Performance Indicators
KYC	Know Your Customer
LMS	Learning Management System
MIS	Management Information Systems
ML	Machine Learning
MLOps	Machine Learning Operations
MOOC	Massive Open Online Course
NLG	Natural Language Generation
NLP	Natural Language Processing
OCR	Optical Character Recognition
OTP	On Time Payment
P2P	Procure to Pay
PAM	Privileged Access Management
PII	Personal Identifiable Information
PoC	Proof of Concept
PoV	Proof of Value
PTO	Personal Time Off
ROI	Return on Investment
RPA	Robotic Process Automation
S & P	Standard and Poor
SaaS	Software-as-a-Service
SCA	Sustainable Competitive Advantage
SCORM	Sharable Content Object Reference Model
SEIM	Security Information and Event Management
SLA	Service Level Agreement
SME	Subject Matter Expert
SOP	Standard Operating Procedures
TQM	Total Quality Management
UI	User Interface
VDIs	Virtual Desktop Infrastructure
VMs	Virtual Machines
WIMP	Windows, Icons+A1:B74, Menus, Pointer

References

2021 Workplace Learning Report. (2021). https://learning.linkedin.com/resources/workplace-learning-report-2021/download-report/3qc; https://learning.linkedin.com/content/dam/me/business/en-us/amp/learning-solutions/images/wlr21/pdf/LinkedIn-Learning-Workplace-Learning-Report-2021-UK-Edition-.pdf.

Agile vs Waterfall: What Is the Difference? Which Is Best for You? (2021, September 30). Trust Radius. https://www.trustradius.com/buyer-blog/difference-between-agile-vs-waterfall.

Alexander Graham Bell Quote: "One Day There Will Be a Telephone in Every Major City in the USA." (n.d.). https://quotefancy.com/quote/765433/Alexander-Graham-Bell-One-day-there-will-be-a-telephone-in-every-major-city-in-the-USA.

Alexe, A. (2018, October 6). *UiPath CEO Daniel Dines: "If Bill Gates Said 'a Computer on Every Desk', Our Vision Is to Have One Robot for Every Person."* Business Review. https://business-review.eu/tech/uipath-ceo-daniel-dines-if-bill-gates-said-a-computer-on-every-desk-our-vision-is-to-have-one-robot-for-every-person-186784.

Analyst, R., Saikat, C. T., Kerremans, M., & Miers, D. (n.d.). *Move beyond RPA to Deliver Hyperautomation.* https://infracom.govt.nz/assets/Uploads/Preparing-for-Technological-Change-in-the-Infrastructure-Sector.pdf. Retrieved February 18, 2021, from https://emtemp.gcom.cloud/ngw/global-assets/en/doc/documents/433853-move-beyond-rpa-to-deliverhyperautomation.pdf.

Anderson, D., & Anderson, L. A. (2010). *Beyond Change Management: How to Achieve Breakthrough Results Through Conscious Change Leadership.* John Wiley & Sons.

Ariola, W. (2021, December 20). *Forrester's Surprising Discovery about Robotic Process Automation.* The New Stack. https://thenewstack.io/forresters-surprising-discovery-about-robotic-process-automation/.

Bae, H. (2015, April 5). *Bill Gates Emails Microsoft Employees to Celebrate Company's 40th Anniversary.* CNNMoney. https://money.cnn.com/2015/04/05/technology/bill-gates-email-microsoft-40-anniversary/index.html.

Bankar, S., & Gankar, S. (2012). Employee Engagement and Change Management. *Journal of Commerce and Management Thought, 4*(2), 313–321. https://www.indianjournals.com/ijor.aspx?target=ijor:jcmt&volume=4&issue=2&article=008.

Beyond RPA: Build Your Technology Portfolio for Hyperautomation. (n.d.). Gartner. https://www.gartner.com/en/webinars/4013885/beyond-rpa-build-your-technology-portfolio-for-hyperautomation.

Brown, S. (2021, April 21). *Machine Learning, Explained.* MIT Sloan. https://mit-sloan.mit.edu/ideas-made-to-matter/machine-learning-explained.

Brownlee. (2020, August 14). *What Is Deep Learning?* Machine Learning Mastery. https://machinelearningmastery.com/what-is-deep-learning/.

Business Wire. (n.d.). https://www.businesswire.com/news/home/20200212005226/en/RPA-Reality-Check-New-Forrester-Research-Identifies-Barriers-to-RPA-Scalability.

Caindec, K. (2021, October 27). *De-Risk Your Digital Transformation Strategy.* Farallon Technology Group. https://www.farallontech.com/2021/10/27/de-risk-your-digital-transformation-strategy/.

Carden, L., Maldonado, T., Brace, C., & Myers, M. (2019). Robotics process automation at TECHSERV: An implementation case study. *Journal of Information Technology Teaching Cases, 9*(2), 72–79.

CBS News. (2022, April 15). *Companies Look to 4-Day Workweek to Help Reduce Stress among Employees without Reducing Productivity.* https://www.cbsnews.com/news/four-day-work-week-california-stress-productivity/.

Chamorro-Premuzic, T. (2021, November 23). The Essential Components of Digital Transformation. *Harvard Business Review.* https://hbr.org/2021/11/the-essential-components-of-digital-transformation.

Chen, H., Liang, Q., Feng, C., & Zhang, Y. (2021). Why and when do employees become more proactive under humble leaders? The roles of psychological need satisfaction and Chinese traditionality. *Journal of Organizational Change Management, 34*(5), 1076–1095.

Clair, C. L. (2021, November 8). *Predictions 2022: The Pandemic's Wake Drives Automation Trends.* Forrester. https://www.forrester.com/blogs/predictions-2022-the-pandemics-wake-drives-automation-trends/.

Colbert, A., Yee, N., & George, G. (2016). The digital workforce and the workplace of the future. *Academy of Management Journal, 59*(3), 731–739.

Companies with a Digitally Savvy Top Management Team Perform Better. (n.d.). MIT CISR. Retrieved October 20, 2022, from https://cisr.mit.edu/publication/2020_0301_TMTDigitalSavvy_WeillWoernerShah.

Davenport, T. H. (2019). What Process Mining Is, and Why Companies Should Do It. *Harvard Business Review.* hbr.org/2019/04/what-process-mining-is-and-why-companies-should-do-it.

Davenport, T. H., & Spanyi, A. (2021, December 2). The Essential Components of Digital Transformation. *Harvard Business Review.* https://hbr.org/2021/11/the-essential-components-of-digital-transformation.

Definition of Advanced Analytics - Gartner Information Technology Glossary. (n.d.). Gartner. https://www.gartner.com/en/information-technology/glossary/advanced-analytics.

Definition of Blockchain - Gartner Information Technology Glossary. (n.d.). Gartner. https://www.gartner.com/en/information-technology/glossary/blockchain.

Definition of Internet of Things (iot) - Gartner Information Technology Glossary. (n.d.). Gartner. https://www.gartner.com/en/information-technology/glossary/internet-of-things.

Definition of Process Mining - Gartner Information Technology Glossary. (n.d.). Gartner. https://www.gartner.com/en/information-technology/glossary/process-mining.

Deloitte editor. (2020, August 19). *Beyond Bots: Scaling Intelligent Automation.* WSJ. Retrieved October 27, 2022, from https://deloitte.wsj.com/articles/beyond-bots-scaling-intelligent-automation-01597863730.

Feijao, C. (2021, December 15). *The Global Digital Skills Gap: Current Trends and Future Directions.* RAND. https://www.rand.org/pubs/research_reports/RRA1533-1.html

Forth, P., Reichert, T., De Laubier, R., & Chakraborty, S. (2020, October 29). *Flipping the Odds of Digital Transformation Success.* BCG Global. https://www.bcg.com/publications/2020/increasing-odds-of-success-in-digital-transformation.

Future Jobs: These are the Fastest Growing and Fastest Declining Roles. (2023, May 1). World Economic Forum. https://www.weforum.org/agenda/2023/04/future-jobs-2023-fastest-growing-decline/

Homepage Wil van der Aalst. (n.d.). https://www.padsweb.rwth-aachen.de/wvdaalst/.

How to Ensure Robotic Process Automation Security. (n.d.). Gartner. https://www.gartner.com/smarterwithgartner/4-steps-to-ensure-robotic-process-auto-mation-security.

Hunkar, D. (2021, May 17). *The Largest Companies by Market Value Change over Time | TopForeignStocks.com.* Retrieved October 26, 2022, from https://topforeignstocks.com/2021/05/17/the-largest-companies-by-market-value-change-over-time/.

IDC FutureScape: Worldwide Digital Transformation 2021 Predictions. (2021). IDC. https://www.idc.com/getdoc.jsp?containerId=US46880818.

Joseph, L., & Clair, C. (2019, July 10). *The RPA Services Market Will Grow to Reach $2 Billion.* Forrester. https://www.forrester.com/report/The-RPA-Services-Market-Will-Grow-To-Reach-12-Billion-By-2023/RES156255.

Kanhonou, C. (2022, March 24). *2021 Gartner Market Guide for Process Mining.* Livejourney. https://www.livejourney.com/gartner-market-guide-2021-pro-cess-mining/.

Khiyara, S. (2018, May 25). *Countdown to a Digital Workforce - Shail Khiyara.* Medium. https://medium.com/@ShailKhiyara/countdown-to-a-digital-workforce-2060b4566712.

Khiyara, S. (2022, November 30). *The Unfulfilled Promise of Automation: DNA Matters.* CIO. https://www.cio.com/article/308690/the-unfulfilled-promise-of-automation-dna-matters.html.

Kosmopoulos, C. (n.d.). *Task Capture vs Process Mining: Everything You Need to Know.* https://www.blueprintsys.com/blog/rpa/task-capture-vs-process-mining-everything-need-to-know.

Learning RPA - Automation Courses | UiPath. (n.d.). UiPath Academy. https://academy.uipath.com/.

London Business School. (2022). *Five Leadership Skills for the Future*. https://www.london.edu/think/five-leadership-skills-for-the-future.

Maldonado, T., Vera, D., & Ramos, N. (2018). How Humble Is Your Company Culture? And, Why Does It Matter? *Business Horizons, 61*(5), 745–753. https://doi.org/10.1016/j.bushor.2018.05.005.

Manchester, H., & Cope, G. (2019). Learning to Be a Smart Citizen. *Oxford Review of Education, 45*(2), 224–241. https://doi.org/10.1080/03054985.2018.1552582.

Marciniak, P., & Stanisławski, R. (2021). Internal Determinants in the Field of RPA Technology Implementation on the Example of Selected Companies in the Context of Industry 4.0 Assumptions. *Information, 12*(6), 222. https://doi.org/10.3390/info12060222.

Markets and Markets (2022). Call Center AI Market with Covid-19 Impact Analysis, By Component, Mode of Channel (Phone, Social Media, & Chat), Application (Workforce Optimization & Predictive Call Routing), Deployment Mode, Vertical and Region - Global Forecast to 2027 (ID: 5589965).

Nacimiento, A. (2019, November 6). *Digital Transformation: Five Practices That Maximize the Chance of Extraordinary Outcomes - By McKinsey's Jacques Bughin, Jonathan Deakin, and Barbara O'Beirne*. Strategic Services. Retrieved October 20, 2022, from https://stratserv.co/2019/11/digital-transformation-five-practices-that-maximize-the-chance-of-extraordinary-outcomes-by-mckinseys-jacques-bughin-jonathan-deakin-and-barbara-obeirne/.

Phillips, D., & Collins, E. (2019). Automation – It does involve people. *Business Information Review*. https://doi.org/10.1177/0266382119863870

Project Management Institute (2017). *A Guide to the Project Management Body of Knowledge (PMBOK Guide)* (6th edn). Newton Square, PA: Project Management Institute.

Reengineering the Recruitment Process. (2021, February 17). Harvard Business Review. https://hbr.org/2021/03/reengineering-the-recruitment-process.

Reskilling Revolution: Preparing 1 Billion People for Tomorrow's Economy (2023, June 27). World Economic Forum. https://www.weforum.org/impact/reskilling-revolution/

Robocorp Survey: The State of RPA 2022. (n.d.). Robocorp. https://robocorp.com/blog/the-state-of-rpa-2022.

Roe, D. (2021, February 16). *How COVID-19 Is Speeding up Change Process in the Digital Workplace*. CMSWire.com. https://www.cmswire.com/digital-workplace/how-covid-19-is-speeding-up-change-process-in-digital-workplace/.

Rogers, D. L. (2016). *Digital Transformation Playbook: Rethink Your Business for the Digital Age*. Columbia Business School Publishing: New York.

Salesforce. (2022, January 27). *Salesforce Launches Global Digital Skills Index: In-Depth Insights from 23,000 Workers*. Salesforce News. https://www.salesforce.com/news/stories/salesforce-digital-skills-index-details-major-gaps-across-19-countries/.

Schroeder, T. (2019, November 4). *BPM vs. Case Management: Understanding the Differences*. SoftExpert Excellence Blog. https://blog.softexpert.com/en/bpm-case-management/.

Schwab, K. (2017). *The Fourth Industrial Revolution*. Currency.

Seasongood S (2016) *Not Just for the Assembly Line: A Case for Robotics in Accounting and Finance*. Financial Executive. Available at: https://www.financialexecutives.org/Topics/ Technology/Not-Just-for-the-Assembly-Line-A-Case-forRobotic.aspx (accessed 19 August 2018).

SGSubra, V. A. P. B. (2022, January 15). *The Essential Components of Digital Transformation*. R2G. https://sgsubra.wordpress.com/2022/01/17/the-essential-components-of-digital-transformation/.

Sharma, S. (2022, September 17). *What Is Conversational AI?* NVIDIA Blog. https://blogs.nvidia.com/blog/2021/02/25/what-is-conversational-ai/.

Solis, B. (2022, February 9). *From the Great Resignation to the Great Digital Skills Divide: The Need for Empathetic Leaders in a Digital-First World*. Forbes. Retrieved October 20, 2022, from https://www.forbes.com/sites/briansolis/2022/02/09/from-the-great-resignation-to-the-great-digital-skills-divide-the-rise-of-empathetic-leaders-in-a-digital-first-world/.

Sonenshein, S., & Dholakia, U. (2012). Explaining Employee Engagement with Strategic Change Implementation: A Meaning-Making Approach. *Organization Science, 23*(1), 1–23. https://doi.org/10.1287/orsc.1110.0651.

Statista. (2021, August 27). *Average Company Lifespan of S&P 500 Companies 1965-2030*. https://www.statista.com/statistics/1259275/average-company-lifespan/#:%7E:text=In%202020%2C%20the%20average%20lifespan,even%20further%20throughout%20the%202020s.

Statista. (2022, May 23). *Nominal GDP Driven by Digitally Transformed and Other Enterprises Worldwide 2018-2023*. https://www.statista.com/statistics/1134766/nominal-gdp-driven-by-digitally-transformed-enterprises/.

The Future of Jobs Report 2020. (n.d.). World Economic Forum. https://www.weforum.org/reports/the-future-of-jobs-report-2020.

The Great Resignation Didn't Start with the Pandemic. (2022, March 25). Harvard Business Review. https://hbr.org/2022/03/the-great-resignation-didnt-start-with-the-pandemic.

The Largest Companies by Market Value Change over Time. (2021). https://topforeignstocks.com/2021/05/17/the-largest-companies-by-market-value-change-over-time/; https://topforeignstocks.com/2021/05/17/the-largest-companies-by-market-value-change-over-time/.

The Top 25 Stocks in the S&P 500. (2022). https://www.investopedia.com/ask/answers/08/find-stocks-in-sp500.asp; https://www.investopedia.com/ask/answers/08/find-stocks-in-sp500.asp.

These 5 Charts Show the Jobs of Tomorrow and the Skills You Need. (2020, October 22). World Economic Forum. https://www.weforum.org/agenda/2020/10/x-charts-showing-the-jobs-of-a-post-pandemic-future-and-the-skills-you-need-to-get-them/.

Training Industry, Inc. (2020, October 27). *The 70-20-10 Model and Experiential Learning in a Multigenerational Workforce*. Training Industry. https://trainingindustry.com/articles/content-development/the-70-20-10-model-and-experiential-learning-in-a-multigenerational-workforce/.

Wikipedia contributors. (2023). Interactive voice response. *Wikipedia*. https://en.wikipedia.org/wiki/Interactive_voice_response

Willcocks, L. P., & Lacity, M. C. (2016). *Service automation robots and the future of work*. https://eprints.lse.ac.uk/87821/

William, J. (2021, February 10). *How to Bring a Culture of Automation into Your Organization*. Forbes. https://www.forbes.com/sites/forbesbusiness-council/2021/02/10/how-to-bring-a-culture-of-automation-into-your-organization/?sh=6dcc2607c656.

World Economic Forum 2021 Upskilling for Shared Prosperity REPORT. (2021). In World Economic Forum. https://www3.weforum.org/docs/WEF_Upskilling_for_Shared_Prosperity_2021.pdf.

World Economic Outlook Databases, October 2020. (2020). IMF. https://www.imf.org/en/Publications/WEO/weo-database/2020/October/.

World Manufacturing Report 2020. (2020, August 4). World Manufacturing Foundation. https://worldmanufacturing.org/activities/report-2020/.

Yse. (2019, January 15). *Your Guide to Natural Language Processing (NLP)*. Towards Data Science. https://towardsdatascience.com/your-guide-to-natural-language-processing-nlp-48ea2511f6e1.

Index

Note: **Bold** page numbers refer to tables and *italic* page numbers refer to figures.